T0135631

Proceedings

of the

International Beilstein Workshop

The Chemical Theatre

of

Biological Systems

May 24th - 28th, 2004

Bozen, Italy

Edited by Martin G. Hicks and Carsten Kettner

BEILSTEIN-INSTITUT ZUR FÖRDERUNG DER CHEMISCHEN WISSENSCHAFTEN

Trakehner Str. 7 – 9
60487 Frankfurt
Germany

Telephone:	+49 (0)69 7167 3211	**E-Mail:**	info@beilstein-institut.de
Fax:	+49 (0)69 7167 3219	**Web-Page:**	www.beilstein-institut.de

IMPRESSUM

Bibliographic information published by *Die Deutsche Bibliothek*

Die Deutsche Bibliothek lists this publication in the *Deutsche Nationalbibliografie*; detailed bibliographic data is available in the internet at http://dnb.ddb.de

ISBN 3-8325-1019-2

Layout by:	Beilstein-Institut	Printed by:	Logos Verlag Berlin
			Comeniushof, Gubener Str. 47
Cover Illustration: Joelle Heyer			10243 Berlin
	Beilstein GmbH		Tel: +49 (0) 30 42 85 10 90
			Fax: +49 (0) 30 42 85 10 92
			Internet: http://www.logos-verlag.de

PREFACE

The Beilstein workshops address contemporary issues in the chemical and related sciences by employing an interdisciplinary approach. Scientists from a wide range of areas - often outside chemistry - are invited to present aspects of their work for discussion with the aim of not only to advance science, but also, to enhance interdisciplinary communication.

To set the stage for the workshop, it is useful to consider the development of both natural and life sciences from their early origins in natural philosophy. Technical equipment and methodologies, as well as, the systematizing and cataloguing of phenomena and entities, have always underpinned scientific advances. However, even in science, there can be resistance to change, and it has often taken a generation of overwhelming experimental evidence to swing opinion, and allow new paradigms to be accepted into the collective scientific wisdom. Whilst technology and information are the driving forces for advances, it is interesting to note that the most significant developments often take place at the intersections of different lines of thought.

In the natural sciences the search for a "life-force" has given way to the generalization that biology can be defined as being interdependent "complicated chemistry". To gain the insights that lead to the understanding of complex processes, the usual scientific method is to break down the problem into smaller units, create a model for each of them, and through refinement of the models attempt to develop a unified theory. Whereas initial insight into biological systems can be obtained by modelling the chemistry of the parts of the system, the properties and functions of the components of a biological system are not those of discrete molecular entities; they are dependent on the presence or absence of other components and their behaviours in relation to one another. Thus modelling the system as a whole is a very complicated if not a highly complex task.

One of the most current challenging problems of the natural and life sciences is the understanding and prediction of the biological chemistry of the cell, with particular reference to the role of organic compounds therein. These molecules are the products of highly refined *in-vivo* and *in-vitro* organic syntheses; they have complex biological functions - making up the systems themselves as well as interacting with and perturbing them. It is our belief that advances can only realistically be achieved in an interdisciplinary environment, where the lines of thought of different scientific cultures are related sufficiently to each other that given the right circumstances, interactions can take place and new developments can follow.

Preface

By raising the curtain on the *Chemical Theatre of Biological Systems* and through the performances of players invited from the areas of chemical, biological and information sciences, our aim is that this workshop, supported by the active participation of the audience, will afford new insights into contemporary scientific issues.

We would like to thank particularly the authors who provided us with written versions of the papers that they presented. Special thanks go to all those involved with the preparation and organization of the workshop, to the chairmen who piloted us successfully through the sessions, and to the speakers and participants for their contribution in making this workshop a success.

Frankfurt/Main, July 2005   Martin G. Hicks
                                             Carsten Kettner

CONTENTS

EVOLUTION OR REVOLUTION:
THE CHALLENGE TO TODAY'S MEDICINAL CHEMIST

STEVEN V. LEY*, IAN R. BAXENDALE AND REBECCA M. MYERS

Department of Chemistry, University of Cambridge, Lensfield Road
Cambridge, CB2 1EW, U.K.

E-Mail: *svl1000@cam.ac.uk

Received: 15th June 2004 / Published: 22nd July 2005

ABSTRACT

As the global emphasis towards more eco-efficient and sustainable practices unfolds before us, so does the new remit for chemistry. We are already applying the principles of this new paradigm to environmentally cleaner and more efficient chemical processes, products and services. This discussion looks more deeply into some of the ways this remit will lead to the evolution of new tools for the molecule maker and how it is poised to revolutionize the way in which synthesis chemists will conduct their programmes in the future.

INTRODUCTION

The modern world of drug discovery is a rapidly changing landscape, and despite the new knowledge resulting from the genome, together with advances in high-throughput screening, informatics and automation, we are seeing increasing research and development expenditure with fewer new drugs actually making it to the market place. Consequently the demands made on the modern medicinal chemists are substantial. We can no longer simply depend upon the evolutionary incremental approach to new molecule discovery: we need to revolutionize our thinking. We must recognize and respond to the greater challenges posed by synthesis in general if chemical processes are to have a sustainable future. Strategic planning must incorporate enhanced productivity as well as environmental considerations.

For these reasons we must have far cleaner reaction processes with significantly improved atom efficiencies and this will require the discovery of many more strategically important reactions and the generation of many more new catalytic processes.

We are also required to make these discoveries more rapidly than ever before yet be mindful of costs and downstream processing. We need much greater diversity in not just the molecules we make but in the chemistry and reagents used to create them. We also have to be much better at using compound design tools especially if we upgrade our thinking to go beyond the molecule - towards supramolecular structures. For this we will need to have improved data-mining and knowledge-capture tools than are currently available. The use of internet trading and other innovations will also have a significant impact. The sixty million known compounds are only a drop in the ocean of what we could create if we were to exploit chemical space to the full - it only takes a quick scan of the therapeutic drugs currently on the shelf to realize we have a long way to go.

When they are first discovered in research labs, new healing drugs are generally synthesized in around a dozen steps and utilize up to fourteen different general types of reaction to construct the required chemical architectures. Although this may convey a somewhat limited spectrum of innovation in drug discovery it is not the full picture as, in fact, the techniques and methods for making molecules *have* greatly improved in recent years [1]. We have witnessed the tremendous speed of technological advances and the impact they have had in every aspect of our lives - drug discovery does not escape this. For example the area of combinatorial *bio*-chemistry, encompassing processes such as gene shuffling, phage display, the use of multiple enzymes and directed evolution techniques, is poised to impact considerably upon the molecule maker's tool box (Fig. 1). Many new chemistry devices such as microarrays, calorimetry, flow reactors, mini-reactor wells and microfluidic systems are becoming routine, and the new generation tools for chemists are increasingly aligning with the nano-technological fascination. Novel solvent systems are becoming popular too, such as ionic liquids, supercritical CO_2, H_2O and the use of fluorous phase materials (Fig. 1). Just as wireless devices are revolutionizing the home and workplace they are also impacting on laboratory management and experimental control and these tools are now considered essential to the drug discovery processes.

There is a growing requirement for synthesis automation systems utilizing design of experiment (DoE) and ReactArray software [2] in association with all manner of robotics, computational tools and high-speed chemical manipulation techniques.

This area is expanding at a phenomenal rate (Fig. 1) [3].

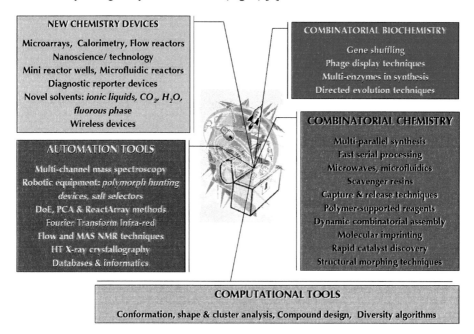

NEW CHEMISTRY DEVICES

Microarrays, Calorimetry, Flow reactors
Nanoscience/ technology
Mini reactor wells, Microfluidic reactors
Diagnostic reporter devices
Novel solvents: *ionic liquids, CO$_2$, H$_2$O, fluorous phase*
Wireless devices

COMBINATORIAL BIOCHEMISTRY

Gene shuffling
Phage display techniques
Multi-enzymes in synthesis
Directed evolution techniques

AUTOMATION TOOLS

Multi-channel mass spectroscopy
Robotic equipment: *polymorph hunting devices, salt selectors*
DoE, PCA & ReactArray methods
Fourier Transform Infra-red
Flow and MAS NMR techniques
HT X-ray crystallography
Databases & informatics

COMBINATORIAL CHEMISTRY

Multi-parallel synthesis
Fast serial processing
Microwaves, microfluidics
Scavenger resins
Capture & release techniques
Polymer-supported reagents
Dynamic combinatorial assembly
Molecular imprinting
Rapid catalyst discovery
Structural morphing techniques

COMPUTATIONAL TOOLS
Conformation, shape & cluster analysis, Compound design, Diversity algorithms

Figure 1. New tools in the molecule maker's tool box.

In order to conduct effective synthesis one needs to recognize that chemistry is part of the continuum from the initial conception and use in the research environment - through the process lab - and on to full-scale production. New synthesis tools should demonstrate applicability across these broad disciplines, even though there can be significant differences in the individual needs and requirements for the chemists involved; speed, safety, scale and cost factors for example.

Over recent years combinatorial chemistry has provided many new opportunities for compound preparation using both parallel processing and fast serial reactions. With the increasing understanding, development of additional practical skills and use of new tools in chemistry such as microwave radiation [4] and microfluidic devices [5], these opportunities will continue to multiply.

Combinatorial chemistry has triggered a renaissance in resin-based synthesis using polymers to support substrates, reagents or scavengers, all of which have aided product work-up and isolation immensely (Fig. 2). In addition, new methods in dynamic assembly of molecules and structural morphing techniques are fast becoming increasingly important drivers for this area of science.

(a) *The simplest case*

(b) *The more complex case*

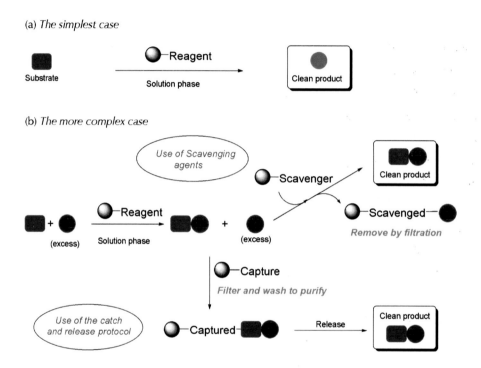

Figure 2. Solid-supported reagents in synthesis. (**a**) The simple case where no by-products are generated. (**b**) A more complex case where excess coupling components have been used or by-product removal is needed.

In spite of some successes arising from the use of combinatorial chemistry it has also been responsible for a certain dumbing down of chemistry by its tendency to rely on straightforward and reliable reactions with accepted concomitant compromises in both yield and complexity of the structures that are synthesized. This is no longer acceptable and we must make compounds designed for purpose rather than simply making compounds because we can. In a research laboratory for example, there might be a requirement for greater diversity in the molecules that are made and high attrition rates may be acceptable. In the process environment however, greater reaction versatility and reliability are priorities thus lower attrition rates are essential.

For these many reasons our group has focused on immobilized reagents, scavengers and catch-and-release techniques as a better practical approach for making chemical compounds, either in a library format or being capable of rapid reaction optimization and eventual scale-up [6].

SUPPORTED-REAGENTS IN SYNTHESIS

Although immobilized reagents have been known since 1946, we saw several advantages and opportunities of this approach when applied to multi-step synthesis programmes. By combining the power of supported reagents, enzymes and other methods such as immobilized scavengers, quenching agents or the catch-and-release principle, tremendous opportunities present themselves. Obviously these methods are well suited to parallel synthesis because of the work-up procedures that simply involve filtration of spent reagents and evaporation to give pure products [7]. The processes are readily automated and the reactions can be followed in real-time using LC-MS and other solution phase analytical techniques. By feeding back information the reaction could also be self-optimizing. Other advantages are that if toxic or volatile compounds must be used, then by immobilizing them on resins they become benign and far easier to handle. Furthermore, when plagued with by-products, co-running impurities, excess reactants or reagents (where conventional chromatography is not only ineffective but also time-consuming) scavenging or catch-and-release techniques become particularly valuable. Many immobilized reagents are also catalytic, or at least the spent reagent can be readily recovered by filtration and recycled to minimize costs. Scale-up is usually straightforward and in the future one could envisage far greater use being made of these supported systems in flow reactors devices [8].

Immobilized reagents are attractive as they allow the piloting of new synthesis schemes on very small quantities of compound. The ability to investigate sequentially a number of new steps in a synthetic pathway by removing the contaminating spent reagents and by-products in this way (rather than using standard protocols of water-quenching, solvent extraction, drying, evaporation and chromatography at each stage) saves considerable time and materials. Even more exciting opportunities arise when the idea of combining several reagents in a single pot to facilitate multiple transformations is considered. The site isolation of reagents (resulting from their immobilization) means that even otherwise mutually incompatible reagents in solution (e.g. oxidants and reductants) do not react together [9].

Without doubt these methods minimize the use of long-winded conventional procedures and create time for more profitable planning, thinking and innovation in the synthesis process. It should also be recognized that not only are one-pot linear synthesis routes possible, but that one can perform convergent syntheses or batch splitting to maximize product variation (Fig. 3). All these are essential components to good synthetic practice.

In a short article such as this it is not possible to cite all the relevant literature, nor can all the work that we have done in the area be covered thoroughly [10] what follows therefore, constitutes a selection of topics to give a flavour of what can be achieved using these systems.

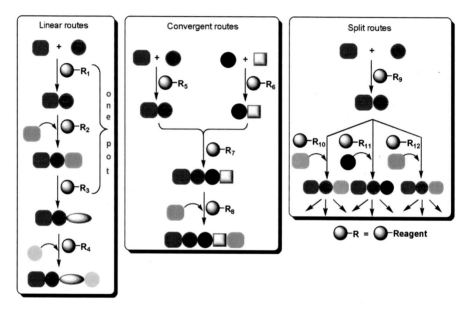

Figure 3. Opportunities for supported reagents.

Firstly, to illustrate the power of these methods we have studied the conversion of a readily available alcohol to an aldehyde using alkylammonium perruthenate resin (Fig. 4) [11]. In this way we generate an aldehyde cleanly by the simple process of filtering away spent reagent. This process can be conducted in real time and monitored using *in situ* methods. The reaction is especially attractive for the generation of unstable aldehydes which are often prone to form hydrates, over oxidize or racemize. The aldehyde thus produced can be used directly in other synthesis programmes by batch-splitting. In this case seven different routes lead to a variety of different products ranging from nitro olefins [12], alkenes and epoxides [13], acids [14], nitriles [15] and enones to isoxazoles [16] (Fig. 4).

All these processes use immobilized reagents and scavengers to effect individual transformations and as a consequence conventional work-up procedures (chromatography, distillation, aqueous washes or crystallization) are unnecessary.

This approach to on-demand synthetic intermediates could be extremely useful in future synthesis programmes as it minimizes waste and compound storage problems particularly as many of the reagents are readily recycled. We have published extensively in this area and suggest the reader consults one of our reviews for further details [6a].

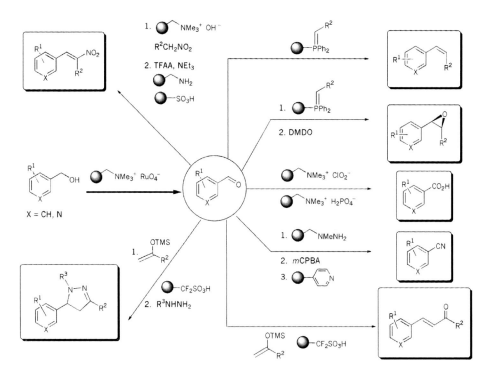

Figure 4. Multiple use of polymer supported reagents by a seven-way split.

MICROENCAPSULATION

Metal-catalysed organic synthesis, especially using palladium, has become common practice, however, product contamination by metal atoms is often problematic and considerable skill and expense are necessary to obtain products with less than 10 ppm of metal contamination.

In order to address this issue we have investigated an approach that effectively entraps the metals within a polymer matrix [17]. Our aim was to generate a system that was inexpensive to operate and would lead directly to clean products by straightforward extraction and a ready recycling of the expensive metal component.

Micro-encapsulation of materials in polymeric coatings is an attractive procedure for drug delivery, radiation therapy, cell entrapment and for the controlled release of pesticides. Microcapsules can be prepared using a variety of chemical and physical methods, and one such method is an *in situ* interfacial polymerization [18]. This involves dispersing an organic phase (containing poly-functional monomers and/or oligomers) into an aqueous phase (containing a mixture of emulsifiers and protective colloid stabilizers) along with the material to be encapsulated.

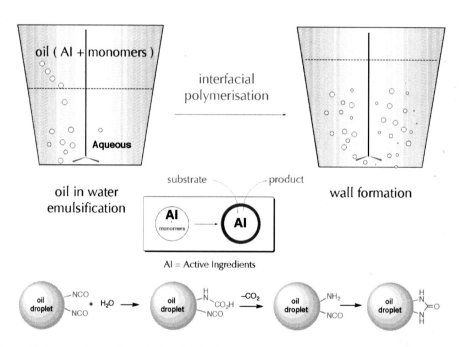

Figure 5. Microcapsule manufacture by interfacial polymerisation.

The resulting oil-in-water emulsion undergoes interfacial polymerization, with the monomers/ oligomers reacting spontaneously at the phase boundary to form microcapsule polymer walls.

The permeability and size of these microcapsules and the coordinating properties of the polymer matrix can be tuned by varying the identity of the monomers/oligomers, the presence of additives and the specific reacting conditions used in the encapsulation such as temperature, concentration or pH. Efficient entrapment of transition metal-based catalysts requires the design of systems possessing ligating functionality in order to retain the metal species. These systems should be physically robust and chemically inert to reaction conditions whilst also being cost effective.

Polyurea microcapsules [19] were found to be suitable by virtue of their chemical structure as they could ligate and retain palladium or other metallic species readily. The micro-encapsulation procedure is straightforward. A solution containing polymethylene polyphenylene diisocyanate (SUPRASEC 5025) and palladium diacetate in dichloroethane was stirred with an aqueous solution of sodium lignosulfonate (Reax 100M), polyvinyl alcohol (Goshenol GL03) and the polyoxypropylene polyoxyethylene ether of butyl alcohol (Tergitol XD) using a standard laboratory overhead stirrer which results in an oil-in-water micro-emulsion. The wall-forming reaction is initiated when some of the peripheral isocyanate groups are hydrolysed at the oil-water interface to form amines; these in turn react with other unhydrolysed isocyanates to form a urea-linked polymeric coating yielding, insoluble and permeable, polyurea microcapsules with a particle distribution of 20-250 microns (average size 150 microns) (Fig. 5). According to X-ray fluorescence (XRF) and inductively coupled plasma (ICP) analysis the average palladium content in the polyurea microcapsules (MC-[Pd]) we made was found to be 0.4 mmol g^{-1}.

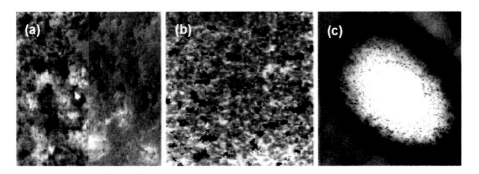

Figure 6. (a) Scanning electron micrograph (SEM) showing the interior of a palladium diacetate containing polyurea microcapsule (magnification 3 x 2500). (b) TEM of a sliced palladium diacetate microcapsule after chemical reaction (magnification 3 x 14.000). (c) TEM of a sliced microcapsule showing the distribution of palladium nanoparticles (dark spots) along a channel fromed within the polyurea matrix (magnification 3 x 46.000).

Our palladium-containing microcapsules, [Pd⁰EnCat], were examined as catalysts in Suzuki cross-coupling reactions of aryl boronic acids with aryl bromides [17], as this is a particularly important transformation. The reactions proceeded extremely well (Fig. 7).

What is important to note is the catalysts were readily recovered by filtration through a polyethylene frit (20 micron). The products of the reaction were examined by ICP analysis and were found to have very low residual palladium content (0.5 - 5 ppm).

Since these early experiments, subsequent reactions have shown the catalysts to be extremely versatile and reliable. They have also found application in super-critical CO_2 in either batch or flow modes [8b, 17b]. To date over 300 different examples of reactions promoted by these catalysts have been conducted. Furthermore, modified catalysts displaying enhanced activity can be made that incorporate additional coordinating ligands, leading to acceleration and tailored selectivity in the reactions.

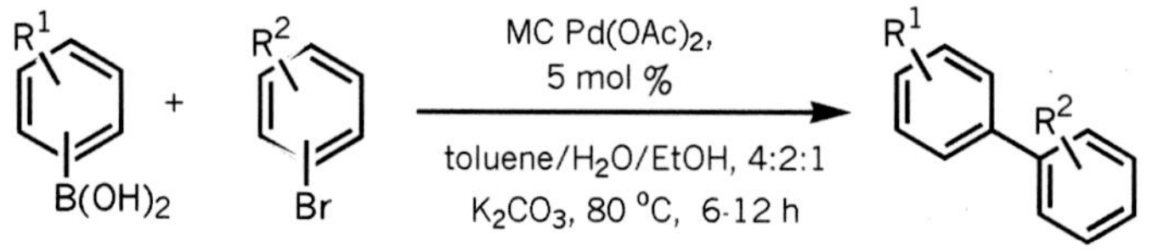

Entry	R^1	R^2	Yield/(%)[†]
1	p-OMe	p-OMe	87
2	p-OMe	p-F	89
3	p-OMe	p-NO₂	91
4	p-OMe	p-OMe	71
5	p-Ac	p-OMe	84
6	p-Ac	p-F	90
7	p-Ac	p-NO₂	97
8	H	p-OMe	94
9	H	p-F	93
10	H	p-NO₂	97

[†](%) based on isolated products. ICP Pd analysis typically 0.5-5 ppm

Figure 7. Synthesis of biaryls using palladium diacetate encapsulated in polyurea.

We have also studied other palladium-catalysed processes such as the Heck and carbonylation reactions with some measure of success (Fig. 8) [17a].

We have demonstrated that these palladium-containing microcapsules function well as catalysts in the hydrogenation of double bonds (Fig. 9) [20].

This important chemical transformation is especially useful when double bond selectivity is required in the presence of other functional features on the molecule which can be sensitive to more commonly used catalysts i.e. palladium on charcoal.

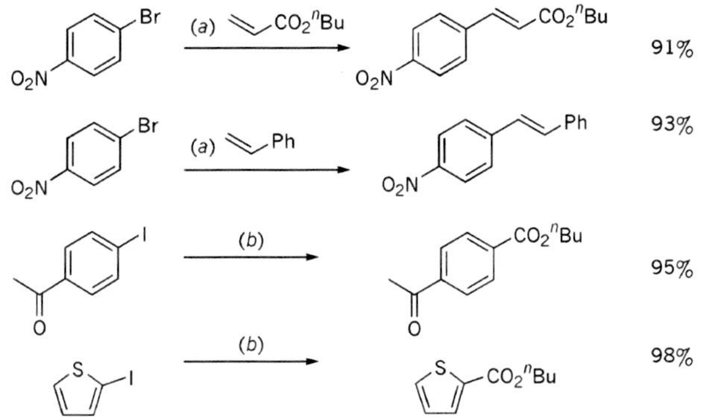

Figure 8. *Conditions*: (**a**) (Heck) 2.5 mol% MC-[Pd], IPA, $^{n}BuN_4OAc$, 90° C plus olefin. (**b**) (Carbonylation) 3 mol% MC-[Pd], CO, $^{n}BuOH$, Et_3N, 90° C.

These new catalysts are recyclable and can be reused at least thirty times. They are also chemoselective; working in the presence of benzyl groups, carbamates, nitriles, epoxides and even aromatic halides. Recovery of the catalyst does not require special filteration procedures nor do we need to wet the catalyst to prevent ignition, as is the case for Pd/C. This safety feature is a significant advantage especially for scale-up applications.

We have continued to develop and evolve these palladium-polyurea catalysts and have extended their application to hydrogen transfer reductions in highly efficient and chemoselective conversions of aryl ketones to the corresponding alcohols (Fig. 10). The most effective conversions were carried out using the encaps with formic acid and triethylamine in the presence of ethyl acetate at room temperature [21].

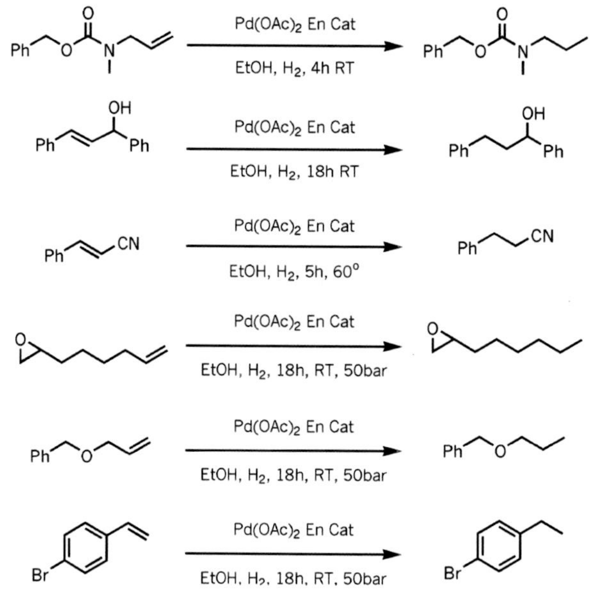

Figure 9. Palladium diacetate En Cats in hydrogenation reactions.

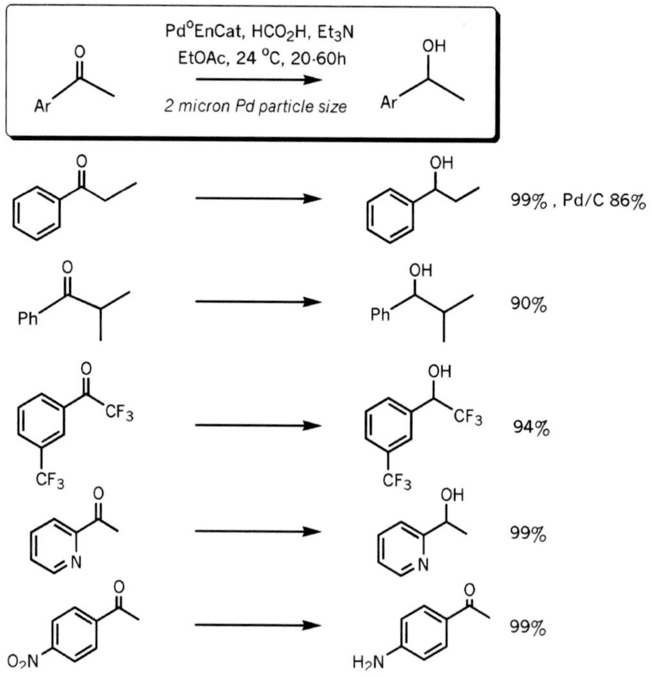

Figure 10. Micro-encapsulated Pd(0) for hydrogenation transfer reduction of aryl ketones.

A further application of this catalytic system is in the reductive ring-opening of benzylic epoxides via transfer hydrogenation (Fig. 11) [22]. Hydrogenolysis of terminal epoxides with conventional hydrogenation conditions was also examined, although this was found to be less effective than the hydrogen transfer process.

It is also noteworthy that these catalysts out-performed palladium/charcoal in any comparison reactions. In conclusion, encapsulated systems are being extended to incorporate many other metals [23], enzymes and even whole cells. With adaptation we can incorporate chiral ligands for asymmetric processes or design cavities suitable for selective scavenging processes. We are also investigating mixed metal perovskites as interesting new catalysts for clean organic synthesis programmes [24].

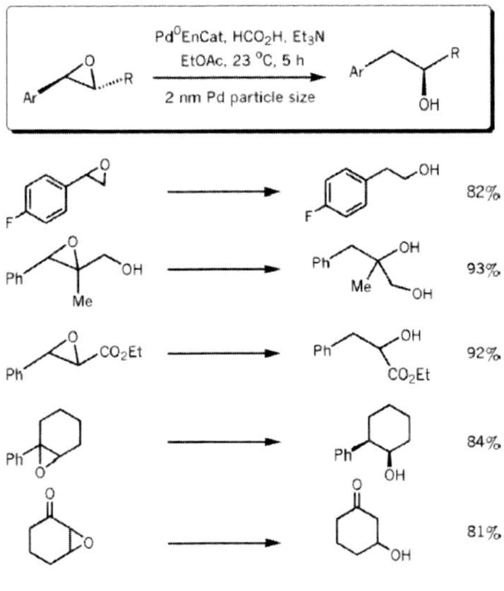

Figure 11. Pd(0) nanoparticles micro-encapsulated for the hydrogenolysis of epoxides.

SUPPORTED REAGENTS IN MULTI-STEP ORGANIC SYNTHESIS

In order to demonstrate the full potential of these immobilized reagents and scavenging agents, we have used them in orchestrated multi-step syntheses of a wide range of building blocks [25] and products such as drug substances [26], azo dyes [27], heterocycles [28] alkaloids and a selection of natural products, some of which are discussed below, highlighting just a few of these applications to demonstrate how useful supported reagents are and how wide ranging their applicability is.

In the first of these projects we wanted to prepare a library of potential anti-proliferative agents histone deacetylase (HDAc) inhibitors in the form of sulfonamide hydroxamic acids [29]. Imbalance in the level of histone acetylation in the body is associated with malignant disease and HDAc inhibitors lead to a reversal of the transcriptional repression and an associated upregulation of tumour suppressors. They have also been observed to result in inhibition of angiogenesis and are therefore of considerable interest as potential new anticancer agents. The sulfonamide hydroxamic acids chosen for study bear a close structural relationship to the natural product *trichostatin A*, a known HDAc inhibitor.

It is also known that the hydroxamic group in these structures could be problematic owing to their propensity to bind metal atoms leading to difficulties in purification; the supported reagent strategy was therefore a logical choice. Accordingly we devised a very efficient set of synthesis protocols for the construction of the three diversity point sulfonamide hydroxamic acid library whereby we were able to use a Zinsser Sofas robotic synthesizer to prepare these compounds with no manual intervention at any intermediate stage (Fig. 12).

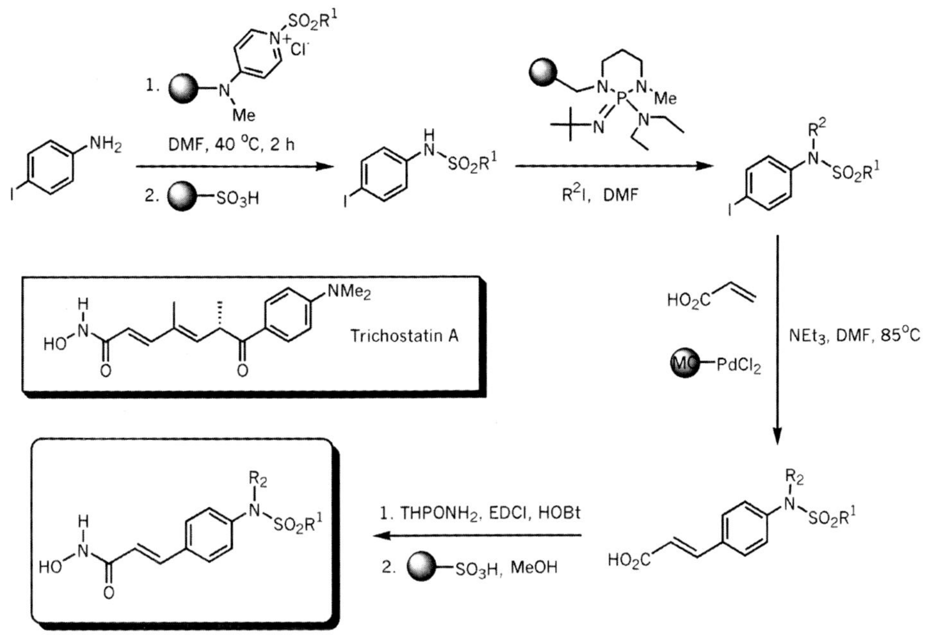

Figure 12. Synthesis of histone deacetylase inhibitors.

Also of importance in this synthesis was that the chosen palladium-encapsulated catalysts for the Heck coupling (involving reaction of the substituted aryl iodides with acrylic acid in DMF) performed especially well. Moreover, in separate experiments with other supported palladium catalysts using a Gilson 233 ReactArray profiling system, the encaps gave a much reduced level of by-products and of course very low levels of palladium contamination in the products. In a similar fashion we have completed the synthesis of a matrix metalloproteinase (MMP) compound collection where once again the final products are hydroxamic acids (Fig. 13) [30]. This synthesis makes use of supported reagents and scavengers in all stages of the process and again leads to very clean products.

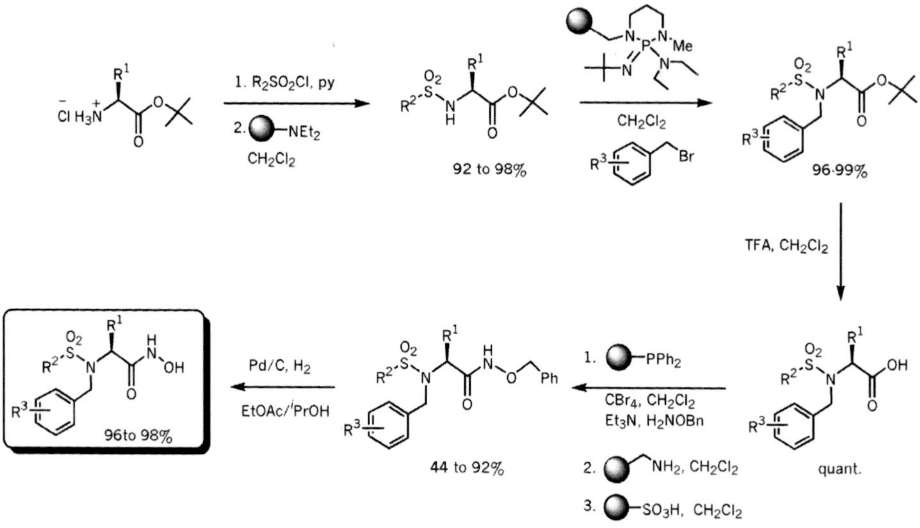

Figure 13. Synthesis of potential matrix metalloproteinase (MMP) inhibitors.

In order to evaluate other supported catalysts for organic synthesis programmes we have carried out initial investigations into how immobilized enzymes would perform during multi-step synthesis operations. This was important as supported enzymes are appealing reagents for synthesis and many of them are already becoming commercially available. Accordingly we have used pig liver esterase (PLE) immobilized on Eupergit® as a key step in the preparation of some γ-amino butyric acid analogues (GABA) (Fig. 14) [31]. This reagent efficiently accomplished the resolution of a cyclopropyl *meso*-diester to a highly enantio-enriched carboxylic acid. This was elaborated through a series of transformations to the corresponding GABA derivative. Noteworthy is the conversion of an aliphatic bromide to a protected amine using an immobilized azide. Direct reductive work-up and Boc-protection was achieved in a single reaction vessel avoiding any need to isolate the potentially hazardous azide intermediates. Although only the cyclopropane ring system has been shown in the scheme, the chemistry translates well to 4, 5 and 6 rings. We believe this application of an immobilized enzyme together with other supported reagents, scavengers and catch-and-release techniques is the first of such sequences to be reported and constitutes a conceptually attractive strategy for complex molecule assembly in the future.

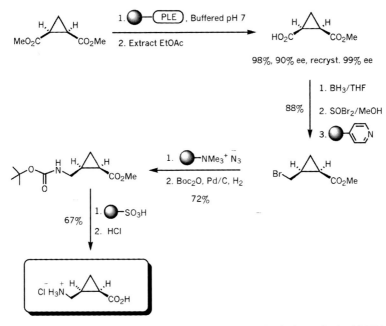

Figure 14. Using immobilised enzymes, in this case pig liver esterase (PLE), in the synthesis of GABA analogues.

The use of these approaches can be illustrated further as in the synthesis of the GSK bronchodilator *salmeterol*, a known β2 agonist used in the treatment of asthma (Fig. 15) [32]. The particular feature of this multi-step process is the convergent strategy and the bringing together two different synthesis streams resulting in a more elaborate, advanced intermediate. This tactic, crucial for the practice of good organic synthesis programmes, is not possible by the more typically used on-bead approach to chemical library generation. During the synthesis of *salmeterol* we also discovered an excellent way of introducing the hydroxymethyl acetal group early on in the synthesis, using a new version of the Mannich reaction. Also of note is the stereoselective introduction of the benzylic hydroxyl group by directed calcium chelation using an immobilized borohydride reagent (Fig. 15).

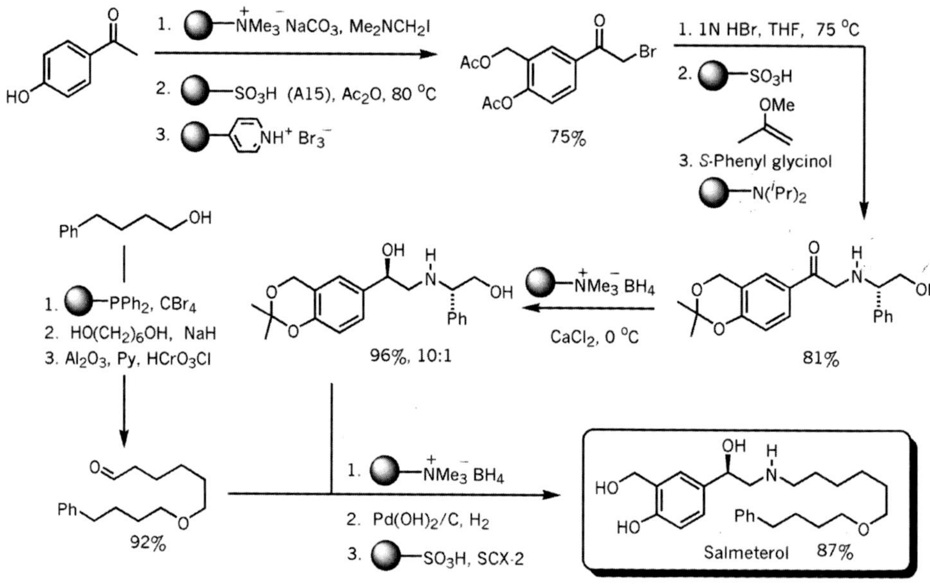

Figure 15. Synthesis of the GSK β2 agonist bronchodilator *salmeterol*.

Salmeterol and related molecules are not always easy to purify by conventional methods owing to their combination of polar heads and lipophilic tails, however the use of supported reagents goes a long way towards overcoming these difficulties. Although we have published the syntheses of several other medicinally relevant molecules including Pfizer's famous erectile dysfunction agent *Sildenafil* (Viagra™) [33], these systems can also be used for complex natural product synthesis [34]. Again we have published extensively in this area, but here we will report only a few selected examples.

NATURAL PRODUCT SYNTHESIS

The synthesis of *carpanone*, a methylenedioxy natural product, constitutes an interesting example as we had to develop new immobilized reagents specifically for the task [35]. Our initial attraction to this particular molecule was the opportunity to effect an oxidative coupling reaction using a substituted phenolic styrene to lead directly to carpanone after an intramolecular Diels-Alder cyclization.

This was achieved using a modified cobalt oxidant as a catalyst in the presence of molecular oxygen giving the natural product in 78% yield as a crystalline material following filtration to remove the spent reagents (Fig. 16).

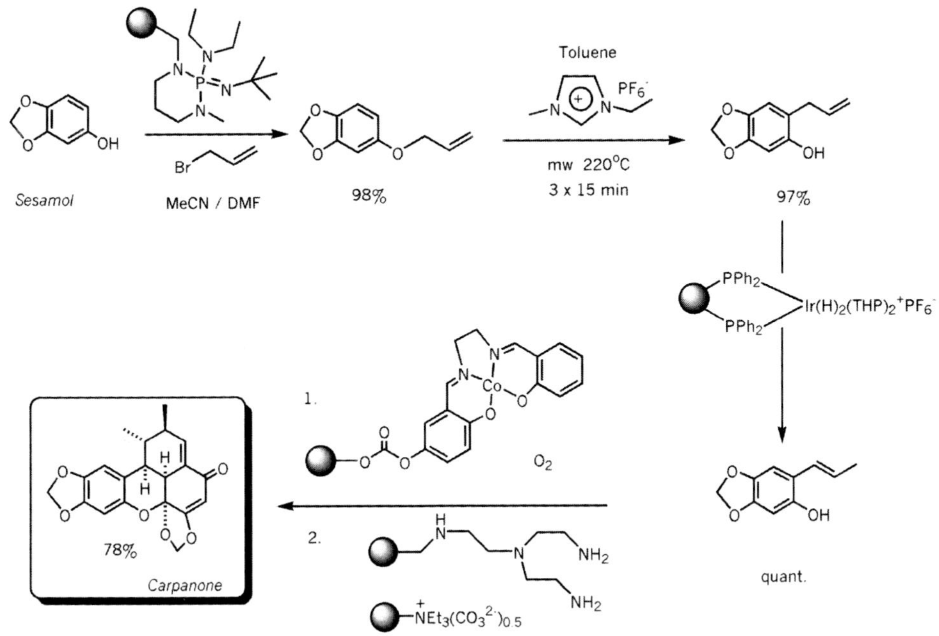

Figure 16. Synthesis of carpanone.

This synthesis neatly demonstrates how simple building blocks can be rapidly transformed into architecturally complex materials. In this work we also invented the use of an immobilized iridium catalyst for the room temperature isomerization of carbon-carbon double bonds, and we have gone on to use it successfully in many other related reactions [36]. Another key step of this synthesis pathway to *carpanone* is the Claisen reaction conducted using a focused microwave reactor (using toluene and an ionic liquid to absorb the microwave energy thus rapidly enabling temperatures of around 200°C). We were one of the early pioneers to use ionic liquids and microwaves in this manner. They were used to effect the smooth conversion of amides to thioamides using a new supported thionlyation catalyst [37] and the first to use immobilized reagents under these conditions.

The next synthesis allowed us to develop extremely efficient methods for alkaloid preparation of some quite challenging structures. Again the emphasis is on ease of operation (with no generation of by-products and reactions amenable to scale-up) using only filtration and evaporation to obtain pure products. A simple six-step synthesis of epimaritidine and oxomaritidine [38] portrays the speed and efficiency of these concepts (Fig. 17).

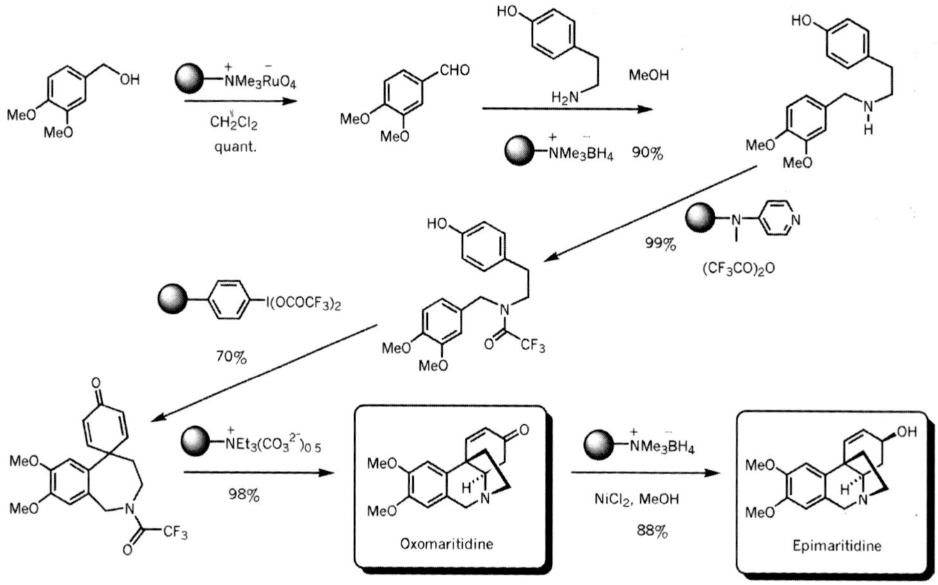

Figure 17. Synthesis of (±)-oxomaritidine and (±)-epimaritidine.

The choice of reagents in the scheme broadly exploits processes we have used for other syntheses. The action of our immobilized perruthenate reagents for example, is both catalytic and recyclable at room temperature and efficiently converts alcohols to aldehydes. Also of more general importance is the polymer-supported hypervalent iodine oxidant [39] used to bring about the phenol oxidative coupling to form the spirodienone product. The last reductive step involving the conversion *oxomaritidine* to *epimaritidine* is new and employs the use of nickel chloride to dope the supported borohydride, presumably to prepare nickel boride *in situ* which then effects the reduction process.

In a more complex example of multi-step natural product synthesis we have prepared the analgesic agent epibatidine [40] (Figs 18-22). This published work makes use of supported reagents, scavengers and catch-and-release techniques.

It also effectively uses microwaves in a key epimerization reaction towards the end of the synthesis, the whole synthesis uses many reagent systems developed by our group. Those skilled in synthesis methods will also appreciate the IRA 420 immobilized base-catalysed Henry reaction leading to nitrostyrenes which are known building blocks for many drug discovery programmes. Once again supported perruthenate and nickel-doped borohydride reagents are exploited to obtain high yield transformations. This synthesis used these reagents in porous polymer pouches to facilitate a more easy removal of the spent reagents on completion of the reaction.

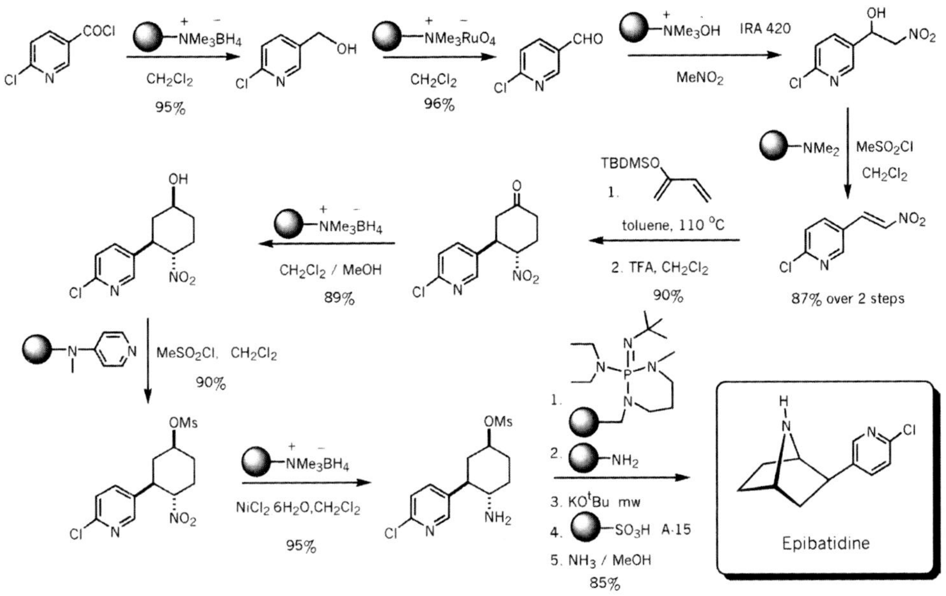

Figure 18. Synthesis of *epibatidine*.

One of the most recent alkaloids we have made using this technology is *plicamine* (Fig. 19) [41]. The route involves a total of fifteen steps carefully designed to afford high quality materials, representing a showcase for the various techniques and immobilized reagents previously described. Once again effective use of the immobilized hypervalent oxidant to give spirodienones is crucial. The use of Nafion, the fluorosulfonylated resin, to bring about the piperidone ring closure is extremely efficient and could be useful in other transformations. Also of note is the new method used to methylate very hindered alcohols with TMS-diazomethane and a sulfonic acid resin, which has many potential applications as a less hazardous alkylation methodology.

The reader is encouraged to consult the full details of this synthesis in which we describe the importance of design of experiment (DoE) and parallel reaction and reagent scanning to optimize many of the reactions. The extensive utilization of focused microwave machinery as described earlier also plays a significant role in this synthesis. One should also recognize that, while this synthesis was targeted at a single molecule, all reaction intermediates can be batch split and diverted into a whole raft of combinatorial chemistry programmes.

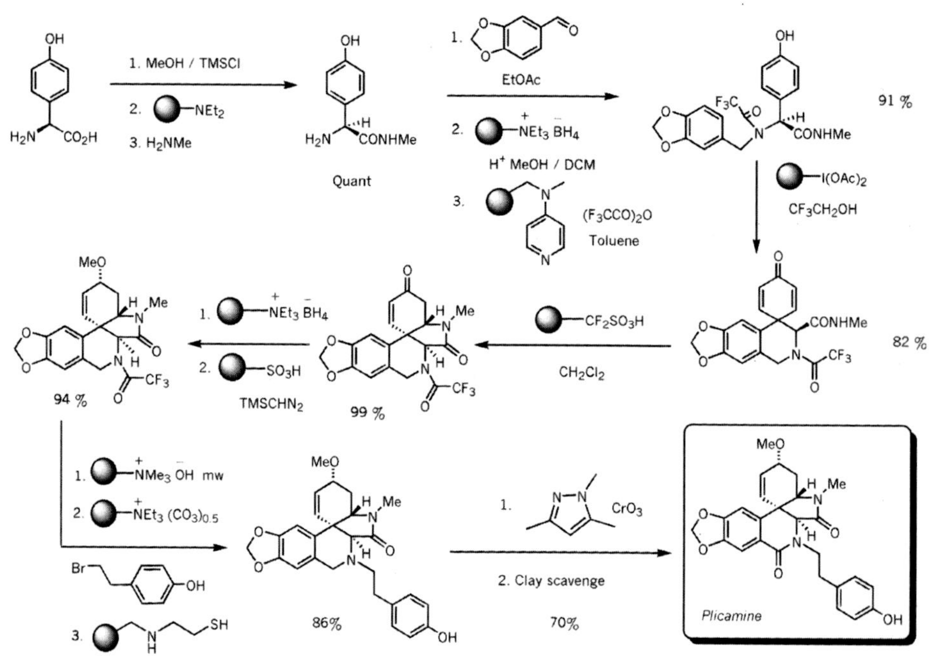

Figure 19. Synthesis of *plicamine*.

Finally the state-of-the-art of these supported reagent methods can be illustrated in the synthesis of the tubulin binder *epothilone C* (Figs 20-23) [42]. The multi-step, multi-convergent approach leads to the natural product more efficiently than any previously described 'in solution' methods. Many of the steps can be optimized rapidly using automated equipment and therefore releases skilled laboratory staff for other tasks. Indeed complex molecule syntheses in the future are likely to be conducted by using a combination of these new tools and techniques especially where routine tasks can be relegated to robotic equipment.

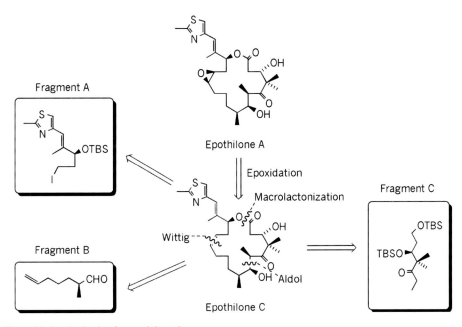

Figure 20. Synthesis plan for *epothilone C*.

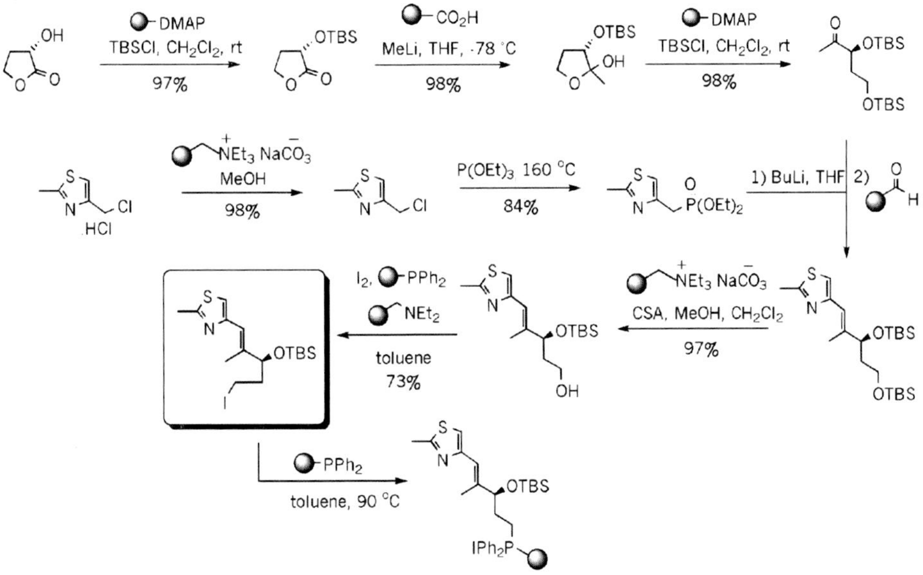

Figure 21. Synthesis of fragment A of *epothilone C*.

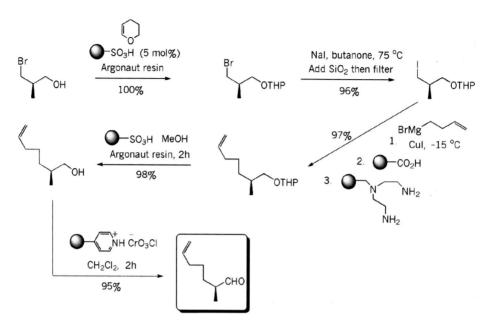

Figure 22. Synthesis of fragment B of *epothilone C*.

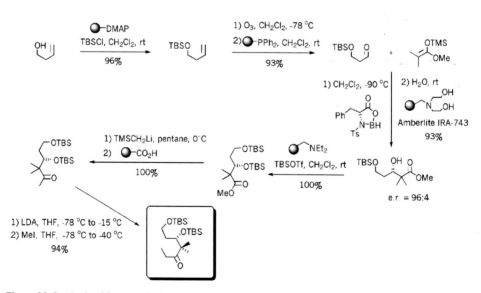

Figure 23. Synthesis of fragment C of *epothilone C*.

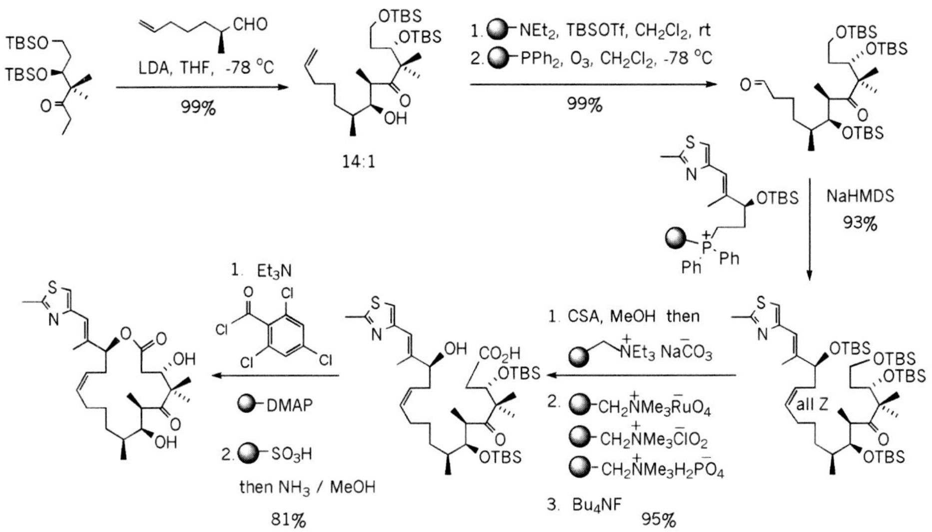

Figure 24. Final steps in the synthesis of epothilone C.

CONCLUSIONS

We remain very excited by the potential of these supported systems for compound synthesis. Future opportunities where stacked or flow reactors can be designed to deliver compounds to order are especially important. Furthermore, the miniaturization of these tools to microfluidic channels or mini-reactor vessels will also be a key driver for future applications. The idea of running one-pot multi-step sequences is now much more feasible. We can envisage using intelligent feed-back mechanisms, the use of robotics and even reagent stirrer bars and other devices to discover new reactions. We have only scratched the surface of synthesis for what will be possible in the years to come [43].

REFERENCES

[1]   (a) Wildonger, R.A., Deegan, T.L., Lee, J.W. (2003) Is combinatorial chemistry on the right track for drug discovery? *J. Autom. Method. Manag. Chem.* **25**:57-61. (b) Nettekoven, M., Thomas, A. W. (2002) Accelerating drug discovery by integrative implementation of laboratory automation in the work flow. *Curr. Med. Chem.* **9**:2179-2190. (c) Seneci, P. (2000) *Solid-Phase Synthesis and Combinatorial Technologies.* John Wiley & Sons. (d) Hird, N.W. (1999) Automated synthesis: new tools for the organic chemist. *Drug Discov. Today* **4**:265-274. (e) Harre, M., Tilstam, U., Weinmann, H. (1999) Breaking the new bottleneck: automated synthesis in chemical process research and development. *Org. Proc. Res. Dev.* **3**:304-318. (f) Van Hijfte, L., Marciniak, G., Froloff, N. (1999) Combinatorial chemistry, automation and molecular diversity: new trends in the pharmaceutical industry. *J. Chromatogr. B.* **725**:3-15.

[2]   (a) Tye, H. (2004) Application of statistical 'design of experiments' methods in drug discovery. *Drug Discov. Today* **9**:485-491. (b) Carlson, R. (1992) *Design and Optimisation in Organic Synthesis.* Elsevier, Amsterdam. (c) Emiabata-Smith, D.F., Crookes, D.L., Owen, M.R. (1999) A practical approach to accelerated process screening and optimization. *Org. Process Res. Dev.* **3**:281-288. (d) Jamieson, C., Congreave, M.S., Emiabata-Smith, D.F., Ley, S.V., Scicinski, J.J. (2002) Application of ReactArray robotics and design of experiments techniques in optimisation of supported reagent chemistry. *Org. Process Res. Dev.* **6**:823-825. (e) Reader, J.C. (2004) Automation in medicinal chemistry. *Curr. Topics Med. Chem.* **4**:671-686. (f) Jamieson, C., Congreave, M.S., Emiabata-Smith D.F., Ley, S.V. (2000) A rapid approach for the optimisation of polymer-supported reagents in synthesis. *Synlett.* 1603-1607.

[3]   (a) Brandli, C., Maiwald, P., Schroer, J. (2003) Automated equipment for high-throughput experimentation. *Chimia* **57**:284-289. (b) Hughes, I., Hunter, D. (2001) Techniques for analysis and purification in high-throughput chemistry. *Curr. Opin. Chem. Biol.* **5**:243-247. (c) Cawse, J. N. (2001) Experimental strategies for combinatorial and high-throughput materials development. *Acc. Chem. Res.* **34**:213-221.

[4]   (a) Baxendale, I.R., Ley, S.V. (2004) Integrated microwave assisted synthesis and solid-supported reagents. In: *Microwave Assisted Organic Synthesis.* (Tierney, J.P., Lidstrom, P., Eds) Blackwell, Oxford. (b) Hayes, B.L. (2002) *Microwaves Synthesis, Chemistry at the Speed of Light.* CEM Publishing. (c) Santagada, V., Perissutti, E., Caliendo G. (2002) The application of microwave irradiation as new convenient synthetic procedure in drug discovery. *Curr. Med. Chem.* **9**:1251-1283. (d) Kappe, C.O. (2002) High-speed combinatorial synthesis utilizing microwave irradiation. *Curr. Opin. Chem. Biol.* **6**:314-320. (e) Larhed, M., Hallberg, A. (2001) Microwave-assisted high-speed chemistry: a new technique in drug discovery. *Drug Discov. Today* **6**:406-416. (f) Caddick, S. (1995) Microwave assisted organic synthesis. *Tetrahedron* **51**:10403-10432. (g) Strauss, C.R., Trainor, R.W. (1995) Developments in microwave-assisted organic chemistry. *Aust. J. Chem.* **48**:1665-1692. (h) Galema, S.A. (1997) Microwave chemistry. *Chem. Soc. Rev.* **26**:233-238. (i) Lidström, R., Tierney, J., Wathey B., Westman, J. (2001) Microwave assisted organic synthesis - a review. *Tetrahedron* **57**:9225-9283. (j) Loupy, A. (Ed.) (2002) *Microwave in Organic Synthesis.* Wiley-VCH, Weinheim.

(k) Blackwell, H.E. (2003) Out of the oil bath and into the oven - microwave-assisted combinatorial chemistry heats up. *Org. Biomol. Chem.* **1**:1251-1255.

[5]   (a) Watts, P., Haswell, S.J. (2003) Microfluidic combinatorial chemistry. *Curr. Opin. Chem. Biol.* **7**:380-387. (b) Fletcher, P.D.I., Haswell, S.J., Pombo-Villar, E., Warrington, B.H., Watts, P., Wong S.Y.F., Zhang, X.L. (2002) Micro reactors: principles and applications in organic synthesis. *Tetrahedron* **58**:4735-4757 and references therein.

[6]   (a) Ley, S.V., Baxendale, I.R. (2002) Organic synthesis in a changing World. *Chem. Record* **2**:377-388. (b) Ley, S.V., Baxendale, I.R. (2002) New tools and concepts in modern organic synthesis. *Nature Reviews* **1**:573-586. (c) Baxendale, I.R., Storer, R.I., Ley, S.V. (2003) Supported reagents and scavengers in multi-step organic synthesis. In: *Polymeric Materials in Organic Synthesis and Catalysis.* (Buchmeiser, M.R., Ed.) Wiley-VCH, Weinheim. (d) Ley, S.V., Baxendale I.R. (2001) Supported catalysts and their applications. The development and application of supported reagents for multi-step organic synthesis. (Sherrington, D.C., Kybett, A.P., Eds) *Royal Society of Chemistry Proceedings* 9-18.

[7]   (a) Ley, S.V., Baxendale, I.R., Bream, R.N., Jackson, P.S., Leach, A.G., Longbottom, D.A., Nesi, M. Scott, J.S., Storer R.I., Taylor, S. J. (2000) Multistep organic synthesis using solid supported reagents and scavengers: a new paradigm in chemical library generation. *J. Chem. Soc., Perkin Trans.* 1. 3815-4195. (b) Kirschning, A., Monenschein, H., Wittenberg, R. (2001) Functionalized polymers - emerging versatile tools for solution-phase chemistry and automated parallel synthesis. *Angew.Chem. Int. Ed.* **40**:650-679. (c) Sherrington, D.C. (2001) Polymer-supported reagents, catalysts, and sorbents: evolution and exploitation - a personalized view. *J. Poly. Sci., Poly. Chem.* **39**:2364-2377. (d) Thompson, L.A. (2000) Recent applications of polymer-supported reagents and scavengers in combinatorial, parallel, or multistep synthesis. *Curr. Opin. Chem. Biol.* **4**:324-337.

[8]   (a) Jas, G., Kirschning, A. (2003) Continuous flow techniques in organic synthesis. *Chem. Eur. J.* **9**:5708-5723. (b) Lee, C.K.Y., Holmes, A.B., Ley, S.V., McConvey, I.F., Al-Duri, B., Leeke, G.A., Santos, R.C.D., Seville, J.P.K. (2005) Efficient batch and continuous flow Suzuki cross coupling reactions under mild conditions. Catalysed by polyurea-encapsulated palladium(II)acetate and tetra-*n*-butylammonium salts. *J. Chem. Soc., Chem. Commun.* (in press).

[9]   (a) Cohen, B.J., Kraus, M.A., Patchornik, A. (1977) Organic synthesis involving multipolymer reactions. Polymeric trityllithium. *J. Am. Chem. Soc.* **99**:4165-4167. (b) Cainelli, G., Contento, M., Manescalchi, F., Regnoli, R. (1980) Polymer-supported phosphonates. Olefins from aldehydes, ketones and dioxolans by means of polymer-supported phosponates. *J. Chem. Soc., Perkin Trans.* 1. 2516-2519. (c) Bessodes, M., Antonakis, K. (1985) One-pot solid-phase cleavage of a-diols to primary alcohols - an attractive route to trihydroxy-nucleosides, antiviral precursors. *Tetrahedron Lett.* **26**:1305-1306.

[10]   (a) Haag, R., Leach, A.G., Ley, S.V., Nettekoven M., Schnaubelt, J. (2001) New polyethylene glycol polymers as ketal protecting groups: a polymer-supported approach to symmetrically substituted spiroketals. *Syn. Commun.* **31**:2965-2977.

(b) Baxendale, I.R., Ley, S.V., Lumeras W., Nesi, M. (2002) Synthesis of trifluoromethyl ketones using polymer-supported reagents. *Comb. Chem. High Throughput Screening* **5**:197-199. (c) Grice, P., Leach, A.G., Ley, S.V., Massi A., Mynett, D.M. (2000) Combined application of analytical techniques for the characterisation of polymer-supported species. *J. Comb. Chem.* **2**:491-495.

[11]   (a) Hinzen, B., Ley, S.V. (1997) Polymer-Supported Perruthenate (PSP): a new oxidant for clean organic synthesis. *J. Chem. Soc., Perkin Trans.* 1. 1907-1908. (b) Hinzen, B., Lenz, R., Ley, S.V. (1998) Polymer-Supported Perruthenate (PSP): clean oxidation of primary alcohols to carbonyl compounds using oxygen as cooxidant. *Synthesis* 977-979.

[12]   Caldarelli, M., Habermann J., Ley, S.V. (1999) Clean five-step synthesis of an array of 1,2,3,4,-tetra-substitued pyrroles using polymer-supported reagents. *J. Chem. Soc., Perkin Trans.* 1. 107-110.

[13]   Bolli M., Ley, S.V. (1998) Developement of a polymer bound Wittig reaction and use in multistep organic synthesis for the overall conversion of alcohols to β-hydroxyamines. *J. Chem. Soc., Perkin Trans.* 1. 2243-2246.

[14]   (a) Yasuda K., Ley, S.V. (2002) The simultaneous use of immobilised reagents for the one-pot conversion of alcohols to carboxylic acids. *J. Chem. Soc., Perkin Trans.* 1. 1024-1025. (b) Takemoto, T., Yasuda, K., Ley, S.V. (2001) Solid-supported reagents for the oxidation of aldelydes to carboxylic acids. *Synlett.* 1555-1566.

[15]   Baxendale, I.R., Ley, S.V., Sneddon, H.P. (2002) A clean conversion of aldehydes to nitriles using a solid-supported hydrazine. *Synlett.* 775-777.

[16]   Haunert, F., Bolli, M., Hinzen B., Ley, S.V. (1998) Clean three step synthesis of 4,5-dihydro-1*H*-pyrazoles starting from alcohols using polymer-supported reagents. *J. Chem. Soc., Perkin Trans.* 1. 2235-2237.

[17]   (a) Ramarao, C., Ley, S.V., Smith, S.C., Shirley, I. M., DeAlmeida, N. (2002) Encapsulation of palladium in polyurea microcapsules. *J. Chem. Soc., Chem. Commun.* 1132-1133. (b) Ley, S.V., Ramarao, C., Gordon, R.S., Holmes, A.B., Morrison, A.J., McConvey, I.F., Shirley, I.M., Smith S.C., Smith, M.D. (2002) Polyurea-encapsulated palladium (II) acetate: a robust and recyclable catalyst for use in conventional and supercritical media. *J. Chem. Soc., Chem. Commun.* 1134-1135.

[18]   Mars, G.J., Scher, H.B. (1990) *Controlled Delivery of Crop Protecting Agents*, (Wilkens, R.M., Ed.), p.65. Taylor and Francis, London.

[19]   Scher, H.B. (1980) U.S. Pat. No. 4,285,720.

[20]   Bremeyer, N., Ley, S.V., Ramarao, C., Shirley I.M., Smith, S.C. (2002) Palladium acetate in polyurea microcapsules: a recoverable and reuasable catalyst for hydrogenations. *Synlett.* 1843-1844.

[21]   Yu, J.Q., Wu, H.C., Ramarao, C., Spencer, J.B., Ley, S.V. (2003) Transfer hydrogenation using recyclable polyurea-encapsulated palladium: efficient and chemoselective reduction of aryl ketones. *Chem. Commun.* 678-679.

[22]   Ley S.V., Mitchell, C., Pears, D., Ramarao, C., Yu, J.Q., Zhou, W.Z. (2003) Recyclable polyurea-microencapsulated Pd(0) nanoparticles: an efficient catalyst for the hydrogenolysis of epoxides. *Org. Lett.* **5**:4665-4668.

[23]   Ley, S.V., Ramarao, C., Lee, A-L., Østergaard, N., Smith S.C., Shirley, I.M. (2003) Microencapsulation of osmium tetroxide in polyurea. *Org. Lett.* **5**:185-187.

[24]   Smith, M.D., Ramarao, C., Brennan, P.E., Stepan A.F., Ley, S.V. (2003) Palladium-containing Perovskites: recoverable and reusable catalysts for Suzuki couplings. *J. Chem. Soc., Chem. Commun.* 2652-2653.

[25]   (a) Hall, B., Haunert, F., Scott, J., Bolli, M., Habermann, J., Hinzen, B., Ley, S.V.,Gervois, A.-G. (1999) Preparation of compounds using polymer supported reagents. Patent WO 9958475. (b) Ley, S.V., Massi, A. (2000) Parallel solution-phase syntheses of functionalised bicyclo-[2.2.2]octanes: generation of a library using orchestrated multi-step sequences of polymer-supported reagents and sequesterants. *J. Chem. Soc., Perkin Trans.* 1. 3645-3654. (c) Congreave, M.S., Kay, C., Scicinski, J.J., Ley, S.V., Williams, G., Murray, P.J., McKeown S.C., Watson, S.P. (2003) Versatile solid-phase synthesis of secondary amines from alcohols. Development of an *N*-Boc-(*o*-nitrobenzene) sulfonamide linker. *Tetrahedron Lett.* **44**:4153-4165. (d) Ley, S.V., Bolli, M., Hinzen, B., Gervois A.-G., Hall, B.J. (1998) Use of polymer-supported reagents for clean multi-step organic synthesis: preparation of amines and amine derivatives from alcohols for use in compound library generation. *J. Chem. Soc., Perkin Trans.* 1. 2239-2241.

[26]   Habermann, J., Ley, S.V., Scott, J.S. (1998) Clean six-step synthesis of a piperidino-thiomorpholine library using polymer-supported reagents. *J. Chem. Soc., Perkin Trans.* 1. 3127-3130.

[27]   Baxendale, I.R., Ley, S.V., Caldarelli M. (2000) Clean and efficient synthesis of azo-dyes using polymer-supported reagents. *Green Chemistry* 43-45.

[28]   (a) Hinzen B., Ley, S.V. (1998) Synthesis of isoxazolidines using Polymer-Supported Perruthenate (PSP). *J. Chem. Soc., Perkin Trans.* 1. 1-2. (b) Habermann, J., Ley, S.V., Smits, R. (1999) Three-step synthesis of an array of substituted benzofurans using polymer-supported reagents. *J. Chem. Soc., Perkin Trans.* 1. 2421-2423. (c) Habermann, J., Ley, S.V., Scicinski, J.S., Scott, J.S., Smits R., Thomas, A.W. (1999) Clean synthesis of a-bromo ketones and their utilization in the synthesis of 2-alkoxy-2,3-dihydro-2-aryl-1,4-benzodioxanes, 2-amino-4-aryl-1,3-thiazoles and piperidino-2-amino-1,3- thiazoles using polymer-supported reagents. *J. Chem. Soc., Perkin Trans.* 1. 2425-2427.

[29]   (a) Bapma, A., Vickerstaffe, E., Warrington, B.H., Ladlow, M., Fan, T.P.D., Ley, S.V. (2004) Polymer-assisted solution phase synthesis and biological screening of histone deacetylase inhibitors. *Org. Biomol. Chem.* **2**:611-620. (b) Vickerstaff, E., Ladlow, M. Ley, S.V., Warrington, B.H. (2003) Fully automated multi-step solution phase synthesis using polymer-supported reagents: preparation of histone deacetylase inhibitors. *Org. Biomol. Chem.* **1**: 2419-2422.

[30]   Caldarelli, M., Habermann J., Ley, S.V. (1999) Synthesis of an array of potential matrix metalloproteinase inhibitors using a sequence of polymer-supported reagents. *Bioorg. Med. Chem. Lett.* **9**:2049-2052.

[31]   Baxendale, I.R., Ernst, M., Krahnert, W.-R., Ley, S.V. (2002) Application of polymer-supported enzymes and reagents in the synthesis of γ-aminobutyric acid (GABA) analogues. *Synlett.* 1641-1644.

[32]   Bream, R.N., Ley, S.V., Procopiou, P.A. (2002) Synthesis of the β_2 agonist (*R*)-salmeterol using a sequence of supported reagents and scavenging agents. *Org. Lett.* 4:3793-3796.

[33]   Baxendale, I.R., Ley, S.V. (2000) Polymer-supported reagents for multi-step organic synthesis: application to the synthesis of Sildenafil. *Bioorg. Med. Chem. Lett.* 10:1983-1986.

[34]   (a) Baxendale, I.R., Brusotti, G., Matsuoka, M., Ley, S.V. (2002) Synthesis of nornicotine, nicotine and other functionalised derivatives using solid-supported reagents and scavengers. *J. Chem. Soc., Perkin Trans.* 1. 143-154. (b) Baxendale, I.R., Davidson, T.D. Ley, S.V., Perni, R.H. (2003) Enantioselective synthesis of the tetrahydrobenzylisoquinoline alkaloid (-)-norarmepavine using polymer-supported reagents. *Heterocycles* 60:2707-2715. (c) Lee, A.-L., Ley, S.V. (2003) The synthesis of the anti-malarial natural product polysphorin and analogues using polymer-supported reagents and scavengers. *Org. Biomol. Chem.* 1:3957-3966.

[35]   (a) Baxendale, I.R., Lee, A.-L., Ley, S.V. (2001) A concise synthesis of the natural product carpanone using solid-supported reagents and scavengers. *Synlett.* 1482-1484. (b) Baxendale, I.R., Lee A.-L., Ley, S.V. (2002) A concise synthesis of carpanone using solid-supported reagents and scavengers. *J.Chem. Soc., Perkin Trans.* 1. 1850-1857.

[36]   Baxendale, I.R., Lee, A.-L., Ley, S.V. (2002) A polymer-supported iridium catalyst for the stereoselective isomerisation of double bonds. *Synlett.* 516-518.

[37]   Ley, S.V., Leach, A.G., Storer, R.I. (2001) A polymer-supported thiolating agent. *J. Chem. Soc., Perkin Trans.* 1. 358-361.

[38]   Ley, S.V., Schucht, O., Thomas, A.W., Murray, P.J. (1999) Synthesis of the alkaloids (±)-oxomaritidine and (±)-epimaritidine using an orchestrated multi-step sequence of polymer-supported reagents. *J. Chem. Soc., Perkin Trans.* 1. 1251-1252.

[39]   Ley, S.V., Thomas, A.W., Finch, H. (1999) Polymer-supported hypervalent iodine reagents in 'clean' organic synthesis with potential applications in combinatorial chemistry. *J. Chem. Soc., Perkin Trans.* 1. 669-671.

[40]   Habermann, J., Ley, S.V., Scott, J.S. (1999) Synthesis of the potent analgesic compound (+)-epibatidine using an orchestrated multi-step sequence of polymer-supported reagents. *J. Chem. Soc., Perkin Trans.* 1. 1253-1255.

[41]   (a) Baxendale, I.R., Ley, S.V., Nesi, M., Piutti C. (2002) Total synthesis of the amaryllidaceae alkaloid (+)-plicamine using solid-supported reagents. *Angew. Chem. Int. Ed.* 41:2194-2197. (b) Baxendale, I.R., Ley, S.V., Piutti, C., Nesi, M. (2002) Total synthesis of the amaryllidacea alkaloid (+)-plicamine using solid-supported reagents. Baxendale. *Tetrahedron* 58:6285-6304.

[42] (a) Storer, R.I., Takemoto, T., Jackson, P.S., Ley, S.V. (2003) A total synthesis of epothilones using solid-supported reagents and scavengers. *Angew. Chem. Int. Ed.* **42**:2521-2525. (b) Storer, R.I., Takemoto, T., Jackson, P.S., Brown, D.S., Baxendale, I.R., Ley, S.V. (2004) Multi-step application of immobilized reagents and scavengers: a total synthesis of epothilone C. *Chem. Eur. J.* **10**:2529-2547.

[43] (a) Ley, S.V., Massi, A., Rodríguez, F., Horwell, D.C., Lewthwaite, R.A., Pritchard, M.C., Reid, A.M. (2001) A new phase-switch method for application in organic synthesis programs employing immobilization through metal-chelated tagging. *Angew. Chem. Int. Ed.* **40**:1053-1055. (b) MacCoss, R.N., Henry, D.J., Brain, C.T., Ley, S.V. (2004) Catalytic polymer-supported potassium thiophenolate in methanol for the removal of ester, amide and thioacetate protecting groups. *Synlett.* 675-678. (c) MacCoss, R.N., Brennan, P.E., Ley, S.V. (2003) Synthesis of carbohydrate derivatives using solid-phase work-up and scavenging techniques. *Org. Biomol. Chem.* **1**:2029-2031. (d) Ley, S.V., Taylor, S.J. (2002) A polymer-supported [1,3,2]oxazaphospholidine for the conversion of isothiocyanates to isocyanides and their subsequent use in an Ugi reaction. *Bioorg. Med. Chem. Lett.* **2**:1813-1816.

THE VALUE OF CHEMICAL GENETICS IN DRUG DISCOVERY

KEITH RUSSELL AND WILLIAM F. MICHNE

Astra Zeneca, CNS Discovery, 1800 Concord Pike, Wilmington 19850-543, U.S.A.
E-Mail: *keith.Russell@astrazeneca.com

Received: 18th August 2004 / Published: 22nd July 2005

INTRODUCTION

To understand what chemical genetics is and how it can add value to the drug discovery process we must first consider some of the challenges and needs in the pharmaceutical industry. The process of discovering new drugs is a highly complex, multidisciplinary activity requiring very large investments of time, intellectual capital, and money. Today the average cost of bringing an NCE to market is on the order of $900 million [1]. For every 5000 compounds synthesized only one makes it to the market. Only three of ten drugs generate revenue that meets or exceeds average R&D costs, and 70% of total returns are generated by only 20% of the products [2]. Given this gloomy backdrop it is even more disturbing to learn that despite the proliferation of many new technologies of great potential (and great cost!) pharmaceutical productivity levels have not increased over the last ten years (which we show graphically in Fig. 1).

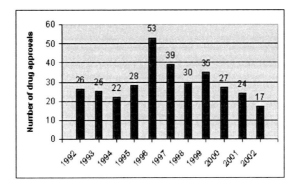

Figure 1. US Drug approvals have not increased over the last ten years.

Pharmaceutical R&D costs continue to grow exponentially, driven in part by investments in new technologies, but the return on this investment remains elusive. There are many reasons for these disturbing trends. If we consider the pharmaceutical industry as primarily a generator of knowledge (defining knowledge as compiled and interpreted information that can be acted upon) and focus on the knowledge creation process, we can shed some light on how the current situation, a productivity gap, emerged. Working harder is not likely to overcome this productivity gap to deliver more drugs. Working smarter, doing things differently, and focusing on what we actually need to deliver, i.e., *knowledge* may be a new way to approach the problem. Ultimately, spanning the "knowledge gap" will lead us to the efficient exploitation of the human genome to discover new drugs to meet major medical needs.

Areas for Improvement

Target Definition - knowing your target

Target Validation - working on the right target

Compound Quality - surviving drug development

Know Early - Build a Strong Foundation

KNOWLEDGE MANGEMENT IN DRUG DISCOVERY

Pharmaceutical companies create and sell knowledge, e.g., knowledge that a drug product will rid patients of the symptoms of their disease while not causing serious side effects. The resources that go into the drug production, pale alongside the resources needed to gain knowledge of what the drug will do when administered to a patient. In the early years of drug discovery it was often true that the literature provided a significant knowledge base for our efforts. Two approaches were taken: 1) function based screening, where one did not know what the target was but could easily screen for small molecules that possessed the right biology [3]; and 2) "rational drug discovery" where one has knowledge of the target and its function [4]. What were needed were small molecules that would interact with the target in the right way before being optimized for *in vivo* activity and safety.

The existing and evolving chemistry and biology literature fuelled these efforts. It is probably also true to say that the medical problems addressed in these early days of drug discovery represented the more accessible opportunities. Often the biology was not only reasonably well understood but it was reasonably easy to study and measure. Examples of biological effects that were tackled include blood pressure, acid secretion and cytotoxicity. The situation today is very different. We now face many new targets that we know little about, and biology that is complex to study and understand. In addition to these issues, advances in our knowledge of distribution, metabolism and pharmacokinetics, as well as toxicology and pharmacogenetics, have led to the introduction of discovery processes that front load measurement of such small molecular properties. This also raises the bar for passage of compounds through to the process - making the process more difficult and slower. While this may lead to lower output of development candidates it should also lead to lower failure rates later in development, i.e., improvements in quality.

KNOWLEDGE GAPS, THEIR IMPORTANCE AND HOW TO ADDRESS THEM

The human genome has been solved and optimistic promises have been made. It is clear that the human genome did not deliver knowledge (i.e., something immediately useful); rather, it delivered a massive amount of data. Significant advances have also been made in cell biology and "systems" biology. The relationship between genes/proteins derived from the human genome and their function as a part of a biological system constitutes the "knowledge gap," and our appreciation for the extent of this void is still emerging. The human genome is thought to consist of ca. 30,000 genes. Each gene can potentially produce several proteins via alternative splicing and post-translational modification, and every protein can potentially combine with other proteins to form many different protein complexes. Clearly, the number of different proteins and protein complexes is much larger than 30,000. o add further complexity, small molecules (that we hope will become drugs) can interact with different sites on a protein or via different mechanisms to further expand the diversity of possible outcomes from the interaction of small molecules with a protein target. We do not know what many gene products (proteins) do, either physiologically or pathologically, and we do not really know how many of these proteins can interact with small molecule ligands [5]. There are many genes about which we know nothing at all.

In summary there is clearly a vast knowledge gap between knowing a gene and knowing the function (physiology and pathology) of its protein product (Fig. 2). The enormity of this knowledge gap has been underestimated by the pharmaceutical industry.

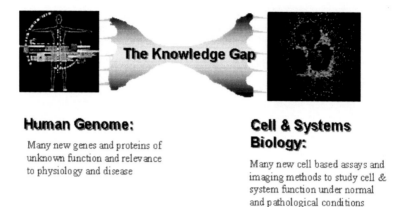

The Knowledge Gap

Human Genome:

Many new genes and proteins of unknown function and relevance to physiology and disease

Cell & Systems Biology:

Many new cell based assays and imaging methods to study cell & system function under normal and pathological conditions

Figure 2. The knowledge gap represents the large gap in understanding that exists between genetic information from the human genome and information regarding biological function from cell and systems biology.

In an effort to illustrate the size of the knowledge gap consider the following (admittedly approximate) analysis from the substance P area. Substance P antagonists have emerged in recent years as potential new treatments for depression although none have yet been approved for this use. Substance P has been known since 1937 and since that time (66 years!) there have been over 6500 papers published providing significant new information on substance P. Thousands of scientists have worked on generating this information over this time frame. It is sobering that our understanding of Substance P's role in depression and in other conditions is still in its infancy! No one pharmaceutical company can generate this volume of information. New, faster and more efficient methods must be developed to fill these knowledge gaps. Partnership with the academic community will become increasingly important as the number of druggable targets expands.

TARGET VALIDATION: THE FOUNDATION OF DRUG DISCOVERY

One critical piece of knowledge to the pharmaceutical industry relates to knowledge of a drug target and its link to a disease process. In the context of small molecule drug discovery we define target validation in a broader sense to include the knowledge of the protein target and its specific interaction with small molecules, and the consequence of this interaction for modifying a disease process. In fact drug discovery is *primarily* focused on the biology of a target in the presence of a drug, i.e., drug induced biology. It begins with a chemical effect - the interaction of a ligand with a protein at a specific site in a specific manner, and ends in patients' gaining benefit from taking a drug derived from the application and exploitation of this knowledge. Target validation that simply links a specific protein and its function to a disease state does not include reference to whether a small molecule can modulate the function of the protein. The protein may not therefore constitute a true "target" since it is not a "target" for a small molecule ligand and efforts to do "target validation" on such a protein will ultimately lead to a negative outcome. We can (and do) proceed to work on drug discovery before we have all the knowledge we need. The absence of this knowledge constitutes the major risk of drug discovery. One way to proceed is to focus on obtaining the most critical knowledge first. This is the knowledge that modulation of a protein target by a small molecule can ultimately lead to a clinical benefit in patients.

CHEMICAL GENETICS - HOW CHEMISTRY CAN CONTRIBUTE TO TI/TV

Target validation is the foundation of drug discovery and needs greater attention if we are to reduce the risk of failure after significant investment. Traditionally target validation has been thought of as a biology problem. Thinking in terms of what knowledge we need makes it clear that the problem does not fall neatly into any particular discipline and is better characterized as an integrated biology and chemistry problem. A schematic target validation "roadmap" is shown in Fig. 3, where the entire validation path from a chemical effect through various levels of biological effects to a clinical effect is outlined.

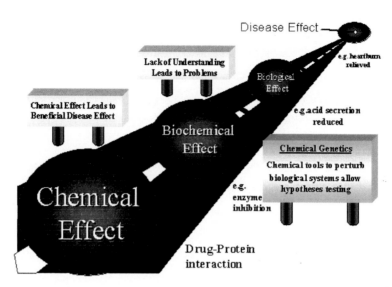

Figure 3. The knowledge road map of target validation beginning with a chemical effect between a small molecule and a protein and ending with a beneficial clinical effect on a human disease. Chemical genetics approaches provide some assistance in pursuing this path.

To begin with, an understanding of the function of a particular gene product in many cases can be achieved through the methods of classical genetics. However, the process can be slow and tedious. For example, developing a mouse carrying the mutation of interest could take months or years. Indeed, if the gene product is essential the organism may not survive long enough to be studied. On the other hand, the situation wherein a molecule is available that alters the function of the gene product has a number of advantages. However, it should be recognized that significant chemical effort is often required. The phenotype of interest is conditional in that it is present only when the molecule is present, allowing the study of essential gene products. It is also tunable, i.e., the intensity of the phenotype can be adjusted by controlling the concentration of the molecule.

> ## <u>What is Chemical Genetics?</u>
>
> **"In Chemical Genetics, small molecules instead of genetic mutations are used to modulate the function of proteins rapidly and conditionally"**
>
> ### *Expanded Definition...*
>
> **Genes can be considered as biological information storage devices**
>
> **Small molecules can be considered as chemical information storage devices.**
>
> *Chemical Genetics is the systematic study of chemotype (the annotated information set that describes a molecule in terms of its interactions with proteins and other macromolecules and the consequences of these interactions.*

Chemical genetics is the purposeful modulation of protein function through interaction with a small molecule. It can also be thought of as the study of chemotype, here defined analogously with phenotype as the annotated information set that describes a molecule in terms of its interactions with proteins and other macromolecules and the consequences of these interactions. The principles of chemical genetics were established in the rich history of using small molecules to explore biological function and in this sense chemical genetics is not new. What is new is the development of a systematic approach to studying biological function with small molecules - this is the emerging field of chemical genetics. Just as genetic changes can alter protein function, so can small molecule-protein interactions [6]. It is important to appreciate that by interaction of a ligand with a protein we mean interaction of a small molecule at a specific site on a protein causing a specific protein change, conformational or otherwise, ultimately leading to a specific biological effect. Small molecules can often interact with multiple sites on proteins and cause a multitude of consequences such as agonism, antagonism, partial agonism, modulation, competitive and non-competitive inhibition, etc.. They can also interact at junctions between protein subunits. The sophistication of small molecule-protein interactions and their biological consequences cannot easily be reproduced by techniques such as gene knock-in/out or using siRNA where genes/proteins are simply removed or increased in concentration in a biological system. Having said that, knockout models have certainly contributed significantly to drug discovery and will continue to do so [7].

The power of chemical genetics resides in this sophistication of the small molecule-protein interaction and the precise way we can (in principle) modulate the function of a protein. As a precursor to drug discovery it serves the purpose to focus us on where small molecule drug discovery really begins - with the chemical interaction of a small molecule and a protein.

The Knowledge We Need

Protein Target	➤ Target Biology
Protein Target + Bound Small Molecule	➤ "Drug+Target" Biology
Small Molecule	➤ Small Molecule Chemotype

At the heart of this approach to knowledge generation in TI/TV (target identification/target validation) is the simple concept that small molecules are used to perturb biological systems. Manipulation of a biological system in a controlled manner by small molecules allows us to study these systems more systematically. In this way the detailed definition of the target - small molecule interaction, and its biological consequence, can be revealed and assessed. This knowledge can be very useful in making decisions about the viability of a drug discovery project.

APPLICATIONS OF CHEMICAL GENETICS: FINDING QUALITY CHEMICAL TOOLS/LEADS

Chemical starting points are needed to develop a lead series and ultimately drug development candidates. They are also needed to help define, validate and aid in the screening of, a biological target where they are often referred to as chemical tools. Finding quality chemical starting points capable of being efficiently optimized into useful tools or drugs is one of the major problems facing the Pharmaceutical Industry. The total number of "reasonable" drug-like molecules has been estimated [8]. The result was approximately 1063 discrete molecules, a number so large that the synthesis of all of them is simply impossible.

LEAD AND CHEMICAL TOOL GENERATION USING INFORMATION RICH COMPOUND SETS

Finding quality chemical tools to modulate biological systems is a difficult step and shares many of the risks associated with finding quality leads in a drug discovery program [9]. Strategies for finding small molecule leads and tools representing two poles on a continuum of approaches are illustrated by structure based design and the high-throughput screening approach. Given our focus on knowledge generation it is interesting to note that molecules at either end of this spectrum also reflect different levels of "information content". Individual molecules used in high-throughput screening teach us (if we are fortunate) about a simple IC_{50} or EC_{50}. Molecules obtained via a structure-based design approach that additionally teach us how they bind to their molecular target, provide us with much more useful information especially when we consider what to do next to improve or change the biology of the molecule (Fig. 4).

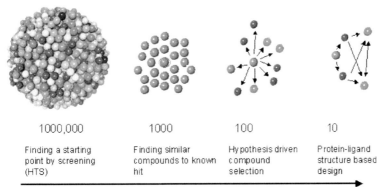

1,000,000	1000	100	10
Finding a starting point by screening (HTS)	Finding similar compounds to known hit	Hypothesis driven compound selection	Protein-ligand structure based design

Increasing information-Better decisions-fewer compounds needed

Figure 4. The spectrum of approaches to finding chemical tools or leads illustrating the inverse relationship between information content and numbers of compounds needed.

Small molecules can reveal other kinds of useful information upon profiling against other targets or biological systems. While knowledge about how a molecule binds to its protein target has been exploited extensively in drug discovery, the profile of information associated with the binding at other non-target proteins has not been fully explored. While some might argue that the outcome of a successful drug discovery project is a small molecule that ONLY interacts with one protein - the target protein - the reality is somewhat different.

IS BIOLOGICAL SELECTIVITY AN ILLUSION?

We have illustrated the enormity of chemistry space and the focus on biologically relevant chemistry space, but what about biology space itself? How many biologically relevant targets are there? While this number has been estimated to be around 3000 [5] it may well be much larger than this if we extrapolate from what we know about particular target classes, e.g., GPCRs, where there are many potential druggable targets and many potential pharmacologies from agonists to antagonists to modulators to inverse agonists. In a typical drug discovery programme selectivity of potential development candidates is often assessed against a panel of 50-100 biologies. Clearly this does not cover a very large fraction of available biology space. In fact many compounds originally thought to be very selective have been found later to have effects against many other targets. For example cholesterol lowering HMG-CoA reductase inhibitors (statins) are among the worlds top selling drugs. It has been recognized recently that statins possess additional biology e.g. anti-inflammatory activity, that is not explained by their interaction with this enzyme. High-throughput screening of large chemical libraries has identified lovastatin (a statin) as an extracellular inhibitor of LFA-1. Lovastatin was shown to decrease LFA-1-mediated leukocyte adhesion to ICAM-1 and T-cell co-stimulation. Unexpectedly, lovastatin was found to bind to a hitherto unknown site in the LFA-1 I (inserted) domain, as documented by nuclear magnetic resonance spectroscopy and crystallography [10].

> **Few, if any, biologically active molecules are highly selective for their "primary" target.**

Some structural classes, e.g. benzodiazepines, are well known to exhibit diverse biology depending on the precise substituent pattern and conformation. Selective ligands with common cores have been obtained against many protein targets (Fig. 5). Such privileged structures suggest that some common structural binding motifs on proteins are reused across many different protein families [11].

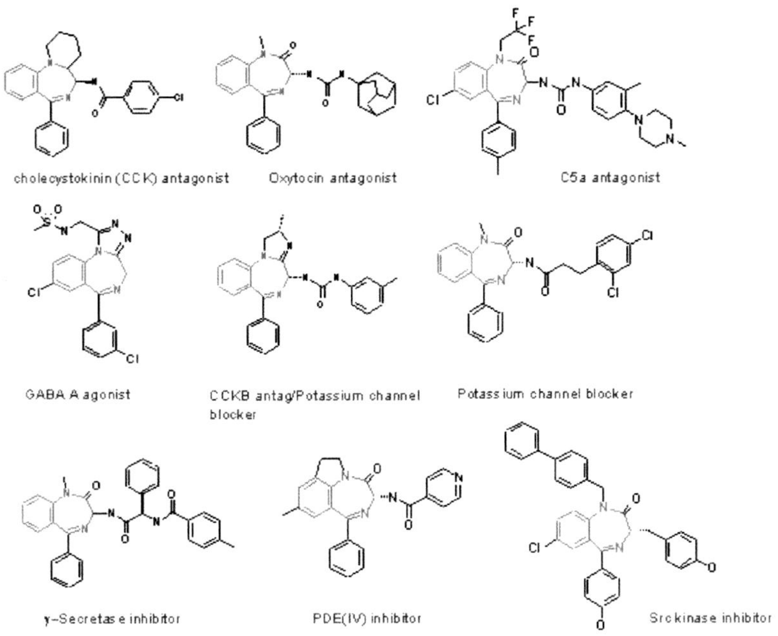

Figure 5. The classic privileged structure, the benzodiazepine nucleus with small structural modification is capable of many different biologies.

It is widely accepted that few if any of the known biologically active molecules are exquisitely selective for a single biological target. This forms the basis for the discovery of new uses for existing drugs and the explanation of side effects observed for all drugs. Indeed, in a commentary on the molecular basis for the binding promiscuity of antagonist drugs LaBella [12] stated that it is unlikely that binding site dimensions, geometry, charge environments, hydrophobic surfaces and other features will ever be known to the extent that drug design technology will yield a compound with absolute specificity for one species of functional protein.

We have used a related strategy to analyse the performance of our corporate collection in high-throughput screening over the past several years (Michne, unpublished results). Our panel of proteins consists of drug targets of interest, and spans several target classes including GPCR's, several classes of enzymes, ion channels, etc.. *Our thesis is that a compound that exhibits biological activity in any target class is more likely to exhibit activity in another unrelated class than is a compound that has never exhibited biological activity of any kind.*

We initially used a relatively small set of assays and screened compounds, and identified about 3500 compounds that were biologically active in at least one assay and met our internal criteria for molecular weight, cLogP, polar surface area, and other chemistry based filters. About 10% of these compounds were found to exhibit activity in other assays. The number of active compounds was then expanded to about 10,000, and the number of assays to 40 (Tinker, unpublished results). The hit rate of the general corporate collection was normalized to a frequency of 1, and compared to the hit rate of the 10K known biologically active set. The results are shown below (Fig. 6).

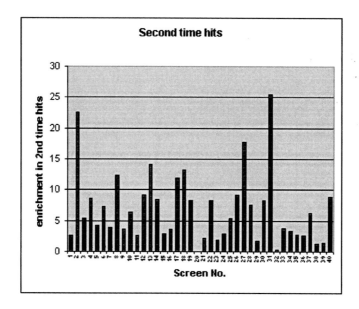

Figure 6. Observed hit rates for a biology based library on a scale where the hit rate for the general collection is normalized to a value of 1.

Clearly, the hit rate exceeds that of the general collection in the majority of screens. However, recent publications sounded a cautionary note. Roche and co-workers [13] reported the development of a virtual screening method for the identification of "frequent hitters." These compounds show up as hits in many different biological assays covering a wide range of targets for two main reasons: a) the activity or the compound is not specific for the target, or b) the compound perturbs the assay of the detection method. They found that with an increasing drug-likeness of the database a decreasing fraction of frequent hitters is predicted.

Sheridan [14] reported finding multi-activity substructures by mining databases of drug-like compounds. Shoichet and co-workers [15] described a common mechanism underlying this phenomenon. In their study they observed that several non-specific inhibitors formed aggregates 30-400 nm in diameter, and that these aggregates were likely to be responsible for the inhibition.

With these two reports in mind we returned to our corporate database and identified, again after suitable filtering, a set of 72,000 biologically active compounds. We then selected a subset of about 25,000 compounds based on the following criteria: a) compounds with confirmed activity in at least two assays; b) compounds with confirmed activity in no more than five assays; c) compounds tested in at least ten assays. We felt that this simple approach would give us a set of information rich compounds largely free of frequent hitters. Using Daylight 2D fingerprints and a Tanimoto distance of 0.3 the set consists of 9,200 clusters, of which there are almost 5,100 singletons. We propose that this richly diverse subset is an ideal starting platform for the design of screening libraries, and for the discovery of new privileged structures. Interestingly, with respect to physical properties, the subset is slightly more lipophilic, and has slightly more polar surface area than the general collection, but the distribution of molecular weights and numbers of hydrogen bond donors and acceptors is the same. We conclude that the currently accepted drug-like physical properties boundary conditions are necessary, but not sufficient to define biological activity, and that other poorly understood factors are the true drivers of such activity. We continue to explore just what those factors might be.

MULTIPARAMETER-ORIENTED COMPOUND SETS

Another approach to the generation of information rich compound sets emphasizes the importance of true integrations of the key disciplines driving drug discovery. Simply put a set of compounds designed to uncover, explore, or provide solutions to key biological, DMPK (Distribution metabolism and pharmacokinetics) and toxicological issues cannot be designed optimally by chemists alone! It is critical that expertize from other disciplines is sought so that the information being used in a medicinal chemistry design campaign is the appropriate information. An important bonus of this focus is that buy-in and teamwork is increased when a group operates in this way (see Fig. 7).

One potential downside of a design strategy that focuses on maximizing the information content revealed by a set of compounds (and often de-emphasizing synthetic accessibility) is that the diversity of structures may present synthetic challenges not easily addressed by typical MPS approaches.

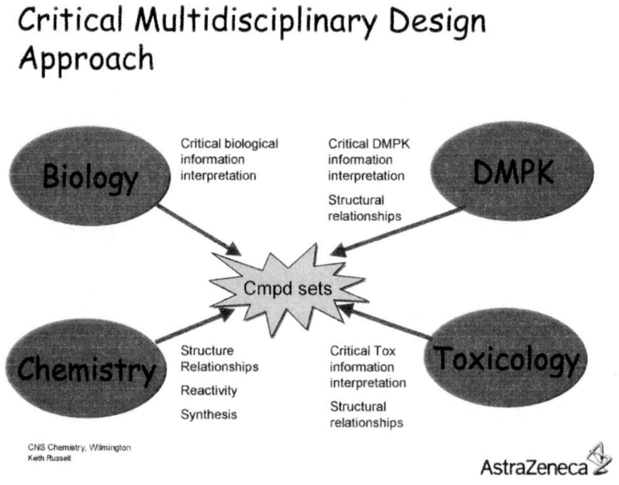

Figure 7. The critical elements of s thorough design process incorporate expertise across many disciplines.

DIVERSITY-ORIENTED SYNTHESIS STRATEGIES: SYNTHESIS OF CHEMICAL GENETICS LIBRARIES

The recognition that the intersection of biology space is limited within chemistry space has encouraged the development of new strategies in organic synthesis for the discovery of biological activity.

Our own interest in this problem was the result of our work on the biology-based collections discussed above. We found that typically only half the compounds were available as solid samples for further study, and that the remainder was dropped from consideration for that reason. The efficient re-synthesis of hundreds or thousands of disparate compounds was simply not practical. Or was it? Perhaps there was an easy way to sort out multiple syntheses to common starting materials and reactions, and carry them out in parallel.

To that end, we used LeadScope software [16] as our management tool. Normally, LeadScope links chemical and biological data, allowing chemists to explore large sets of compounds by a systematic substructural analysis using a predefined set of 27,000 structural features. More importantly for our purposes, two sets can be compared with respect to these features. We chose the Available Chemicals Directory (ACD) as our second set.

We could then easily select those starting materials that would give rise to many products via different routes. We then run as many reactions as possible using parallel synthesis methods. We have used this method for syntheses of up to four steps, and have been able to maintain a productivity level of one compound per chemist per day, 25 mg scale, purified >85%, and characterized by LC-MS and nmr.

We are also developing an approach to the true simultaneous synthesis of disparate core compounds. Most molecules of the size and complexity that we are interested in would likely be prepared in no more than five steps. The actual transformations are usually limited to the chemistry background and experience of the chemist(s) involved in the project. However, the routes need not be so limited. Indeed, consider the generation of tens or hundreds of routes to each compound of interest. The problem then becomes one of how to prepare the maximum number of compounds using the minimum set of common chemistries, staging the routes as necessary in order to maximize the overlap of reagents and conditions. The generation of syntheses is software based. Two or three decades ago there was a lot of effort to develop software to predict the most efficient syntheses of complex organic molecules; most have been abandoned. We chose to use the SynGen program [17] for the very reason that it usually produces several routes to a molecule, each of which begins with a commercially available starting material, and its transformations usually have a literature precedent.

Common chemistries can be grouped at three levels: a) reaction type, e.g., acylation of amines; b) reagent type, e.g., acylation of secondary amines; and c) specific reagents, e.g., acylation of diethyl amine. Each level is specifically encoded by the program, making searching, sorting, and matching fairly easy. We will not necessarily choose the shortest route to each molecule, since it is entirely possible that some longer routes would give rise to additional commonalities thereby allowing the preparation of a larger total number of compounds. As shown in the example a set of 27 diverse compounds were synthesized using this approach in 22 steps.

This is a significant efficiency gain over the 83 synthetic steps needed to access these compounds using a more traditional approach.

Thus greater efficiency is achieved by staging the syntheses to maximize overlap across steps. It is anticipated that larger libraries will result in even greater synthetic efficiency (Fig. 8).

Synthesis Matrix
27 compounds require 22 synthetic operations

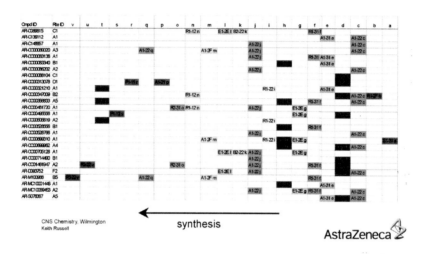

Figure 8. Illustration of a synthesis matrix aligning common synthetic steps aimed at the synthesis of 27 diverse small molecules.

SUMMARY

Bridging the knowledge gap between the data provided by the human genome and our knowledge of biological processes and systems is a requirement for the efficient and effective exploitation of this knowledge in drug discovery. We see this knowledge gap as being best bridged by a truly interdisciplinary approach and a tight integration of chemistry, biology, thinking and experiment. Chemical genetics provides a framework for the systematic study of small molecules to perturb and thus understand biological systems. The adoption of chemical genetics thinking is already growing in its influence among chemists and biologists, and the fruits of this integrated approach to drug discovery promises to be an exciting and rewarding area of research for the next decade.

REFERENCES

[1]   "Post-approval R&D raises total drug development costs to $897 million". (Kaitin, K. I., Ed.), *Tufts Center for the Study of Drug Development Impact Report* 2003, 5(3), can be found under http://csdd.tufts.edu/InfoServices/Publications.asp#ResearchBibliography.

[2]   "Pharmaceutical Innovation: An analysis of leading companies and strategies": *Reuters Business Insight* 2002, can be found under http://www.delphipharma.com/innovation.htm.

[3]   Olbe, L., Carlsson, E., Lindberg, P. (2003) *Nature Rev. Drug Discov.* **2**:132-139.

[4]   Black, J.W.(1989) *Science* **245**:486.

[5]   Hopkins, A.L., Groom, C.R. (2002) *Nature Rev. Drug Discov.* **1**:727-736.

[6]   a) Schreiber, S.L.(2003) *C&EN* 81:51-61; b) Schreiber, S.L.(1998) *Bioorg. Med. Chem.* **6**:1127-1152; c) Stockwell, B.R. (2000) *Trends Biotechnol.* **18**:449-455.

[7]   Zambrowicz, B. P., Sands, A. T. (2003) *Nature Reviews* **2**:38-51.

[8]   a) Bohacek, R.S., McMartin, C., Guida, W.C. (1996) *Med. Res. Rev.* **16**:3-50; b) Kolb, H. C., Finn, M.G., Sharpless, K.B. (2001) *Angew. Chem. Int. Ed.* **40**:2004-2021.

[9]   a) Gura, T. (2000) *Nature* **407**:282-284; b) Waldmann, H. (2002) *Angew. Chem. Int. Ed.* **41**:2879-2890.

[10]   Weitz-Schmidt, G. (2002) *Trends Pharmacol. Sci.* **23**:482-486.

[11]   Holm, L. (1998) *Curr. Opin. Chem. Biol.* **8**: 372-379.

[12]   LaBella, F.S. (1991) *Biochem. Pharmacol.* **42**: Suppl., S1-S8.

[13]   Roche, O., Schneider, P., Zuegge, J., Guba, W., Kansy, M., Alanine, A., Bleicher, K., Danel, F., Gutknecht, P., Rogers-Evans, E., Neidhart, M., Stalder, W., Dillon, H., Sjögren, M., Fotouhi, E., Giillespie, N., Goodnow, R., Harris, W., Jones, P.,Taniguchi, M.,Tsujii, S., von der Saal, W., Zimmerman, G., Schneider, G. (2002) *J. Med. Chem.* **45**: 137-142.

[14]   Sheridan, R.P. (2003) *J. Chem. Inf. Comput. Sci.* **43**:1037-1050.

[15]   McGovern, S.L. Caselli, E. Grigorieff, N. Shoichet, B.K. (2002) *J. Med. Chem.* **45**:1712-1722.

[16]   Roberts, G., Myatt, G.J., Johnson, W.P., Cross, K.P., Blower, P.E. (2000) *J. Chem. Inf. Comput. Sci.* **40**:1302-1314.

[17]   a) Hendrickson, J.B. (1997) *Knowl. Eng. Rev.* **12**:369-386; b) Further information and a demonstration of the program can be found under http://syngen2.chem.brandeis.edu/syngen.html.

CHANGING PARADIGMS IN DRUG DISCOVERY

HUGO KUBINYI

University of Heidelberg, Donnersbergstrasse 9, D-67256 Weisenheim am Sand, Germany

E-Mail: kubinyi@t-online.de

Received: 5th May 2004 / Published: 22nd July 2005

ABSTRACT

The strategies of drug design have changed significantly within the past few decades. Whereas chemistry, biological activity hypotheses and animal experiments dominated drug research, especially in its "golden age", from the sixties to the eighties of the last century, many new technologies have developed over the past 20 years. A vast amount of new drugs was expected to result from combinatorial chemistry and high-throughput screening; however, the yield of new drug candidates was relatively poor. Molecular modelling, virtual screening and 3D structure-based design support the selection and rational design of high-affinity protein ligands. But high affinity to a disease-relevant target is only one important property; in addition, a drug must be orally bioavailable, it should have favourable pharmacokinetics and no unacceptable side effects or toxicity. The following questions are discussed in detail: what are the reasons for the productivity gap between R&D costs and the number of NCE's? Is there a "druggable genome"? Is target focus always best? Is poor ADME the main problem in clinical development? Are we using the right virtual screening tools? What are the main problems in virtual screening and structure-based design? What is wrong and what could we do better?

INTRODUCTION

The strategies of drug research did not change much from the late 19th century until the seventies of the 20th century. New compounds were synthesized and tested on animals or organ preparations, following some chemical or biological hypotheses. Although synthetic output was relatively low, the real bottlenecks were the biological test models. Pharmacological experiments using dozens of animals for every new compound often needed more time for biological characterization than for chemical synthesis.

This situation started to change about 30 years ago. Slowly rational approaches developed, such as QSAR and molecular modelling. The consequence was a lower output in such projects, when certain chemical structures that were proposed by these methods had to be synthesized. On the other hand, *in vitro* test systems such as enzyme inhibition or the displacement of radio-labelled ligands in membrane preparations enabled much faster investigation of new analogues. Now chemistry was the bottleneck. About 10 to 15 years ago, another significant shift in drug discovery paradigms occurred: combinatorial chemistry suddenly flooded the biology laboratories with an overwhelming number of new compounds. It has been commented that combinatorial chemistry was the "revenge of the chemists" to the development of fast *in vitro* test models, with their large output of data within a relatively short time. However, biologists were able to compete: tens of thousands of compounds, later even more, could be investigated by automated high-throughput screening (HTS) systems in just one week.

In the past, wrong or misleading results were obtained too often just because of the use of animal models. Gene technology made an important contribution to drug discovery: the opportunity to produce almost any protein in sufficient quantities enabled biologists to test new compounds at human targets. Genetically modified animals indicate whether a certain principle could work in therapy. The action of an enzyme inhibitor can be simulated before any compounds are synthesized and tested, by a knockout of the corresponding enzyme; the action of drugs can be investigated in animals bearing a human protein. In addition, the production of larger quantities of a protein of therapeutic relevance allows the determination of its three-dimensional (3D) structure at atomic resolution by protein crystallography, alternatively by multidimensional NMR methods. As a consequence, methods developed for the structure-based design of ligands, by modelling or experimental determination of the 3D structures of protein-ligand complexes. Unfortunately, a new bottleneck resulted! Early combinatorial chemistry was guided by synthetic accessibility and the hype for large numbers. Due to this wrong focus, a huge amount of greasy, high-molecular weight compounds resulted, with all their problems in bioavailability and pharmacokinetics. Biological testing which did not produce any valuable hits or supposed hits, later failed in preclinical or clinical development. Whereas this situation has fortunately changed in recent years, due to the maturation of combinatorial chemistry to an automated parallel synthesis of designed libraries, there was still a need for the fast measurement or prediction of ADME (absorption, distribution, metabolism, excretion) properties. Indeed, ADME became the new bottleneck.

Nowadays, we have information on the sequence of the human genome; our combinatorial chemistry approaches are under control by medicinal chemists and biologists; in addition to structure-based design we apply computer-aided methods for data mining, virtual screening, docking and scoring, to predict valuable leads and optimized candidates; HTS models have developed to ultra-HTS models, with up to a million test points per 24 hours; we even have fast experimental models and prediction tools for some ADME parameters. Is there a new bottleneck? Yes, unfortunately, or better to say: yes, of course. Target validation, the proof that the modulation of a certain target by a small molecule will indeed work in therapy, is one of the new bottlenecks. The other one, even more problematic, is the fact that only for some targets can small molecules be discovered which modulate the protein or a certain protein-protein interaction in the desired manner; "druggable" defines the property of a certain target to be accessible by small molecule intervention.

In the past, serendipity played a big role in the discovery of new drugs [1-3]. Some other projects confirmed that the search for new drugs may be more efficient by establishing biological or structure-activity hypotheses and/or selecting certain scaffolds and substituents in the design of new drug candidates. The ratio of 10,000 compounds to produce one new drug is still very often cited. This applies to the situation where research starts from an endogenous ligand or any other lead structure. The "irrational approach", to test huge numbers of in-house compounds, commercially available compounds or chemistry-driven combinatorial libraries in HTS, did not deliver the expected amount [4]. Hundreds of thousands to millions of compounds have to be investigated if such a search starts from scratch, without any knowledge of an active lead, and even then there is no guarantee of success.

THE "DRUGGABLE GENOME"

The human genome project has provided information on all our genes. However, the situation is the same as the one with Egyptian hieroglyphs before the discovery of the Rosetta Stone. We read the text but we understand only a minor part. There are about 30,000 genes in the human genome but we do not know how many of them are disease-related and how many of the gene products will be druggable. It has been estimated that about 600 to 1500 druggable, disease-related targets exist, if one assumes about 10% of disease-related genes on the one hand and about 10% druggable gene products on the other hand [5]. However, this number has to be questioned because only the number of genes was considered [6].

First of all, many proteins or protein variants (estimated to be in the range of some 100,000s) are produced by alternative splicing and/or post-translational modification than there are genes in the genome. Second, proteins can form a multitude of heteromeric complexes that are made up from only a small number of different proteins, e.g. GABA and nicotinic acetylcholine (nACh) receptors, integrins and heterodimeric G protein-coupled receptors (GPCR). Third, some proteins are involved in more than one signalling chain, interacting with different proteins to modulate certain effects. And, last but not least, many therapeutically used drugs do not interact with just one target but have a balanced effect on several different targets. A striking example for such a promiscuous drug is the atypical neuroleptic olanzapine, which interacts as a nanomolar ligand with many different GPCRs [7-9].

Thus, we should neither discuss a druggable genome, nor a druggable proteome, nor a "druggable targetome", but a "druggable physiome" [6]. Our problem is that we do not yet know how to define and design a drug with the right balance of different target affinities, e.g. for depression, schizophrenia and other CNS diseases.

VIRTUAL SCREENING

Several new strategies have been developed for the structure-based and computer-aided design of active compounds. Drug research has often been compared with the search for a needle in a haystack. If neither active leads nor the 3D structure of the biological target are known, HTS seems to be the only reasonable approach. But much useful information can be derived from virtual screening [10], which reduces the size of the haystack. First of all, reactive compounds and other compounds with undesirable groups can be eliminated by so-called "garbage filters" [11]. In a next step, the Lipinski (Pfizer) rule of five may be applied to estimate the potential for oral bioavailability; this set of four rules demands that the molecular weight of a molecule should be lower than 500; the lipophilicity, expressed by log P (P = calculated octanol/water partition coefficient), should be smaller than 5; the number of hydrogen bond donors should not be larger than 5; the sum of oxygen and nitrogen atoms in the molecule (as a rough approximation of the number of hydrogen bond acceptors) should not exceed 10 [12]. A high risk of insufficient oral bioavailability is assumed if more than one of these conditions is violated. Often the rule of five-compatible molecules is erroneously called "drug-like" [6]. However, most of the compounds of the ACD (Available Compounds Directory) [13] would get this label if only the Lipinski rules are applied. "Drug-like" or "non-drug-like" character can only be attributed by neural nets that have been trained with drugs and chemicals [14-16]. In this context it is important to notice that filters are valuable and efficient in the enrichment of interesting candidates out of large libraries. Single compounds should not be evaluated by such filters because the relatively large error rate of about 20% false positives and 20% false negatives would too often provide misleading results.

The situation in drug discovery is much better if a certain number of active and inactive ligands of a target are already known. If a chemical series belongs to a common scaffold or to some related scaffolds, 2D or 3D similarity methods, QSAR and 3D QSAR approaches, and pharmacophore approaches can be applied to derive structure-activity hypotheses (some of the problems of pharmacophore generation will be discussed in the next section). The results of such analyses are proposals for new syntheses or selections of compounds from a library. A highly valuable tool in this respect is feature tree similarity comparisons [17,18], where the molecules are coded as strings with nodes, to which the pharmacophoric properties of the corresponding functional group, ring or linker are attributed. Due to this simple representation of the molecules, similarity searches can be performed extremely fast. In this manner, screening hits can be compared in their similarity to a whole in-house compound library, to libraries of commercially available compounds, e.g. the MDL Screening Compounds Directory [19], and to even larger virtual libraries.

If the 3D structure of a new target is known from experimental determination or from reliable homology modelling, the situation seems to be better but in reality it is not. There remains a high degree of uncertainty about the 3D structure of the protein in the bound state if no information on protein-ligand complex 3D structures is available. Relatively often, the protein itself and its ligand complexes have significantly different 3D structures, the most prominent example being HIV protease. In addition, the relatively low resolution of most protein 3D structures does not allow one to differentiate between the side chain rotamers of asparagine, glutamine, threonine and histidine; the protonation state of histidine remains unclear; water molecules, which are important for the binding of a ligand, are sometimes neglected in protein 3D structures.

All these problems exist only to a minor extent if several protein-ligand complexes can be inspected, which leads to the fourth and best situation in ligand design: not only the protein but also some protein-ligand complexes are known. Molecular modelling and docking aids in the design of new ligands with hopefully improved binding affinity and/or selectivity with respect to other targets.

It should be emphasized that structure-based design can result in a high-affinity ligand but affinity is only a necessary property of a drug, not a sufficient one. In addition, a drug has to be bioavailable, it must have a proper biological half-life time and it must not be toxic, among other important properties.

PHARMACOPHORES

The definition of a pharmacophore is simple [20]. A 3D pharmacophore corresponds to an arrangement of hydrogen bond donor and acceptor, lipophilic and aromatic groups in space, in such a manner that these moieties can interact with a binding site at the target protein; in addition, steric exclusion volumes can be defined. However, the identification of a pharmacophore within a congeneric group of compounds is far from being trivial. Although there are computer programs for the automated derivation of pharmacophores from series of active and less active analogues [21], a better and more reliable method seems to be a "construction by hand" [22]. Four independent problems have to be considered:

- the different pharmacophoric properties of oxygen atoms,
- the protonation and deprotonation of ionizable groups,
- the consideration of tautomeric forms, and
- the superposition of flexible molecules.

Oxygen atoms are strong hydrogen bond acceptors, as long as they are either connected to a carbon atom by a double bond (e.g. in aldehydes, ketones, carboxylic acids, carbonyl group of esters) or substituted by hydrogen and/or aliphatic residues (water, aliphatic alcohols and aliphatic ethers). They are weak or even not acceptors at all (e.g. the sp^3 oxygen atom of an ester group) if their directly neighbouring atoms are connected to another atom by a double bond or if they are part of an aromatic system, as in oxazoles and isoxazoles (Fig. 1) [23,24].

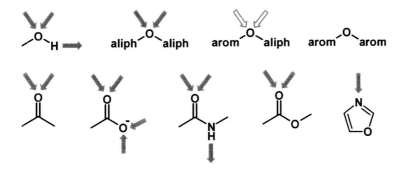

Figure 1. The oxygen atoms of alcohols, aliphatic ethers, aldehydes, ketones, carboxylates and amides are strong hydrogen bond acceptors. The oxygen atoms of mixed aliphatic-aromatic ethers are weaker acceptors and the oxygen atoms of aromatic ethers and heterocycles are more or less without acceptor properties. The same applies to the sp3 atom of an ester group, because of the electron-withdrawing effect of the carbonyl group, which itself is a strong hydrogen bond acceptor, and to oxygen in aromatic systems.

Ionizable groups must be recognized and defined in the right manner to end up with correct pharmacophores. As this is still a mainly unsolved problem for many compounds that are not simple acids, phenols or anilines (at least considering the speed that is needed in the virtual screening of large libraries), a rule-based system has recently been proposed [25]. In this set of rules, all carboxylic acids, the strongly basic amidines and guanidines, and quaternary ammonium compounds are permanently charged. Neutral and protonated forms are generated and investigated in parallel for amines, imidazoles, pyridines and other nitrogen-containing heterocycles. For tetrazoles, thiols, hydroxamic acids, and activated sulfonamides, neutral and deprotonated forms should be investigated in parallel. Certain rules restrict the number of generated species, to avoid combinatorial explosions: there are definitions of the maximum number of charges in a molecule and no identical charges are allowed in adjacent positions of the molecule. Although this approach is definitely better than using all molecules in their neutral form, refined prediction models are urgently needed. An even more difficult problem arises from the fact that ionizable amino acid side chains in proteins may significantly change their pK_a value in dependence of their environment [26,27].

Protomers and tautomers constitute another serious problem in virtual screening and docking (Fig. 2) [28,29]. 1,3-Diketones, acetoacetic esters, hydroxypyridines, oxygen-substituted pyrimidines and purines, and many other compounds may exist in several tautomeric forms that have to be recognized and considered.

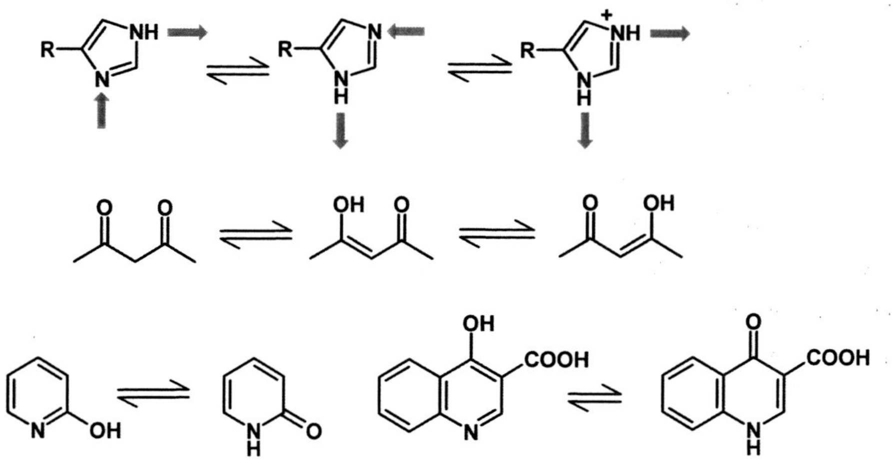

Figure 2. The two different protomers of imidazole (upper left) present their donor and acceptor moieties in different positions; as imidazole has a pK_a value around 7, also the charged form with two donor functions has to be considered (upper right). The other compounds are typical examples of tautomeric forms of molecules, where donor and acceptor functions change their position.

For the purine base guanine, 15 different tautomers can be formulated [29]. In this context it is interesting to note that for a long time Watson and Crick had problems in deriving the correct 3D structure of DNA because they only considered the enol tautomers of guanine and thymine, instead of the keto forms (Fig. 3). When their colleague Donohue corrected this error, they immediately arrived at the correct base pairing [30]. Sometimes enol forms of a ligand are induced by the binding site, as is the case for the binding of the barbiturate Ro 200-1770 to a matrix metalloprotease (Fig. 3) [31]. Computer programs for the generation of all possible tautomers have been described [29,32].

If all these topics are considered in an appropriate manner, the next step is an alignment of the molecules. With rigid molecules, this is most often no problem. But even with steroids the question arises, whether a 3-keto-17-hydroxy-steroid and a 3-hydroxy-17-keto-steroid should be aligned according to their molecular skeleton (which puts the hydrogen bond donor groups of both molecules far apart) or whether a head-to-tail superposition is more favourable; the steric superposition is about as good in this latter case as in the conventional superposition [33]. For flexible molecules, the most rigid active species should be used as a template onto which, step by step, the other molecules are superimposed.

This can be done by visual inspection or by field fit methods, such as SEAL [34,35]; most often several different conformations have to be considered. FlexS is a computer program for flexible superposition; one molecule is used as a rigid template and all other molecules are superimposed in a flexible manner onto this template [36,37]. 3D QSAR methods, such as CoMFA (comparative molecular field analysis) [38,39] or CoMSIA (comparative molecular similarity index analysis) [40], surprisingly do not depend on knowledge of the bioactive conformation. If all conformations are "wrong" to the same extent, the result of an analysis may nevertheless be useful. Another difficulty in the alignment of molecules arises from different binding modes of seemingly similar molecules; there are no general rules on how to recognize such situations.

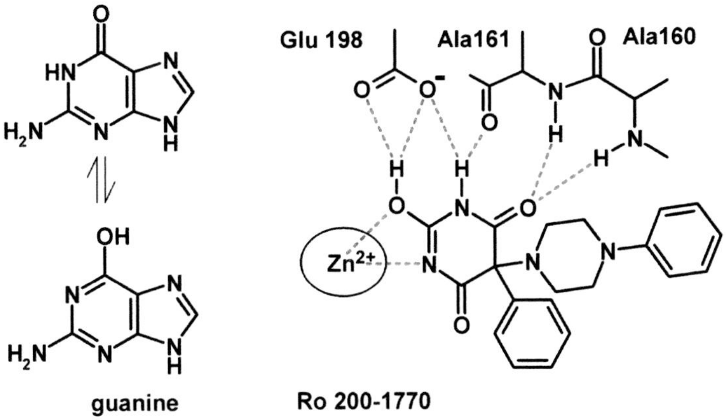

guanine   Ro 200-1770

Figure 3. The guanine tautomer shown in the upper left is the predominant one of 15 possible tautomers, whereas Watson and Crick, worked for a long time, with the tautomer shown in the lower left. Ro 200-1770 is a matrix metalloprotease inhibitor. Only one tautomer can bind with high affinity; the carbonyl form or the other enol form will not form favourable hydrogen bond networks.

Once a pharmacophore hypothesis has been derived, 3D searches can be performed, using commercial software [21,41]. However, it must be emphasized that 3D searches are only meaningful if all the structures of a database are defined according to their correct pharmacophoric properties; otherwise such searches are just useless.

STRUCTURE-BASED LIGAND DESIGN

In the 1970s, the first structure-based design of ligands was performed. The 3D structure of the 2,3-diphosphoglycerate (2,3-DPG) haemoglobin complex was used to derive simple aromatic dialdehydes which mimicked the function of 2,3-DPG as an allosteric effector molecule. Another early example was the structure-based design of trimethoprim analogues with significantly improved affinities to dihydrofolate reductase. However, neither the haemoglobin ligands nor the trimethoprim analogues could be optimized to become drugs for human therapy [42,43]. The first real success story was the structure-based design of the antihypertensive drug captopril, an angiotensin-converting enzyme (ACE) inhibitor. The structure of captopril was derived in a rational manner from a binding site model, using the 3D information of an inhibitor complex of the closely related zinc protease carboxypeptidase A [44].

With the ongoing progress in protein crystallography and multidimensional NMR techniques, the 3D structures of many important proteins, especially enzymes, have been determined. This information led to the structure-based design of several therapeutically useful enzyme inhibitors, most of which are still in preclinical or clinical development. Marketed drugs that have resulted from structure-based design are e.g. the antiglaucoma drug dorzolamide (Merck) and the newer HIV protease inhibitors nelfinavir (Agouron Pharmaceuticals, now Pfizer) and amprenavir (Vertex Pharmaceuticals; developed and marketed by GSK).

Neuraminidase is an interesting target for the structure-based design of anti-influenza drugs. In a very elegant study, Mark von Itzstein used the computer program GRID to estimate interaction energies of the neuraminic acid binding site of this enzyme with different probe atoms or small groups [45]. He realized that the introduction of basic groups, like $-NH_2$, $-C(=NH)NH_2$ or $-NH-C(=NH)NH_2$, into the relatively weak inhibitor neu5ac2en should significantly improve inhibitory activities. This is indeed the case: the neuraminidase inhibitor zanamivir is about 4 orders of magnitude more active than its 4-hydroxy-analogue neu5ac2en. Due to its polar character, zanamivir (Relenza®, GSK) is orally inactive; it must be applied by inhalation [46].

Scientists at Gilead Sciences started from the observation that the glycerol side chain of certain zanamivir analogues does not contribute to affinity. In a series of carbocyclic analogues, strongest inhibitor activity was observed for a pent-3-yl ether. Its ethyl ester prodrug oseltamivir (GS 4104, Tamiflu®, Hoffmann-La Roche; Fig. 4) shows good oral bioavailability [46,47]. Several other success stories of structure-based design have been published [43,48-51].

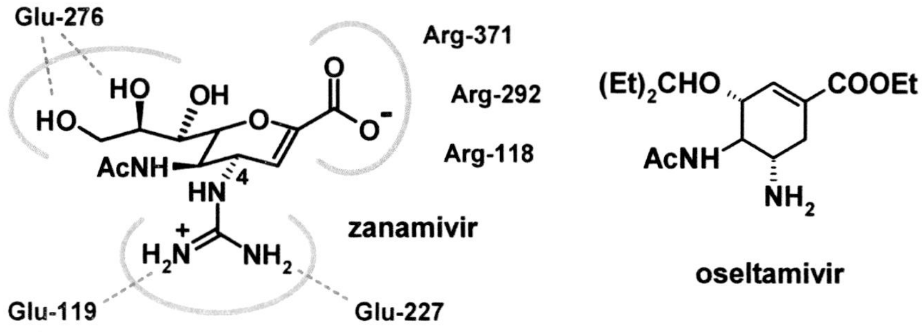

Figure 4. Schematic presentation of the binding mode of the 4-desoxy-4-guanidino-analogue of neu5ac2en, zanamivir, to neuraminidase (left); chemical structure of the orally available prodrug oseltamivir (right).

COMPUTER-AIDED LIGAND DESIGN

Whereas structure-based design can be regarded as the predominant strategy of the last two decades, several computer-assisted methods have been developed more recently. If thousands of candidates and even larger structural databases are to be tested for their suitability to be ligands of a certain binding site, this can no longer be performed by visual inspection. The design process has to be automated with the help of the computer.

The very first computer-based approaches, to search for ligands of a certain binding site, were the programs DOCK [52] and GROW. The *de novo* design program LUDI, developed by Hans-Joachim Böhm at BASF in the early 1990s, was a significant improvement over these early prototypes [53,54]. After the definition of a binding site region by the user, the program automatically identifies all the hydrogen bond donor and acceptor sites, as well as aliphatic and aromatic hydrophobic areas of this part of the protein surface. From the program-implemented information on the geometry of the interaction of such groups with a ligand, the program creates vectors and regions in space, where the complementary groups of a ligand should be located. In the next step, LUDI searches any database of 3D structures of small and medium-sized molecules for potential ligands. Every candidate is tested in a multitude of different orientations and interaction modes, optionally also in different conformations. After a rough evaluation, by counting the number of favourable interactions and by checking for unfavourable van der Waals overlap between the ligand and the protein, the remaining candidates are prioritized by a simple but efficient scoring function [55].

This scoring function estimates interaction energies on the basis of charged and neutral hydrogen bonding energies, hydrophobic contact areas, and the number of rotatable bonds of the ligand. In a last step, the program is able to attach groups, fragments and/or rings to a hit or to an already existing lead structure. The flexible docking of ligands onto a rigid binding site can be achieved by the programs DOCK 4.0 [56], GOLD [57], FlexX [58,59], and the public domain program AutoDock [60,61], to mention those which are the most prominent; more docking programs and several success stories of computer-assisted drug design have been reviewed by Schneider and Böhm [62]. The FlexX modifications FlexE [63] and Flex-Pharm [64] allow a flexible ligand docking into an ensemble of different binding site conformations and the definition of pharmacophore constraints, respectively. Of course, the pharmacophoric properties of all molecules must also be defined in a correct manner in structure-based and computer-aided design.

FRAGMENT-BASED AND COMBINATORIAL LIGAND DESIGN

Several other methods for the design of new ligands have been described in the past, e.g. needle screening [65,66], which starts from a collection of small drug-like ligands and attempts to extend the best ones to larger ligands. In the binding of biotin to avidin, some molecular fragments have only micromolar affinities, whereas biotin itself binds with femtomolar affinity [67]. This principle has recently been used in the rational design of a nanomolar enzyme inhibitor from two low-affinity natural products which bind to different sites of the protein [68]. The SAR by NMR method [69-71] searches for small, low-affinity ligands of proteins which bind to adjacent areas of the binding site. A linker combines both molecules to a nanomolar ligand. Some other NMR-based techniques for ligand discovery have been developed [72-78].

Fragment-based ligand design has been applied for combinatorial techniques [79]. Up to 10,000 low-molecular weight ligands can be tagged onto a gold-coated glass surface [80]; binding of any protein to these microarrays of immobilized ligands is detected by surface plasmon resonance, in this manner avoiding the development of a specific screening method for a new protein. The dynamic assembly of ligands [81-84] generates new molecules from fragments which reversibly react with each other in the presence of a protein.

Molecules that fit the binding site are preferentially formed and afterwards trapped by a reaction which freezes the equilibrium. Some other approaches for the combinatorial design of new leads have been described [85-89].

There are also several computer-assisted techniques for the combinatorial combination of fragments to new leads. The program CombiGen [90] designs libraries with a high percentage of drug-like compounds by assembling privileged and/or user-defined fragments and optionally modifying the resulting structures; virtual screening procedures eliminate molecules with undesired properties. TOPAS [91,92] dissects lead structures into fragments and assembles new molecules by re-combining a chemically similar scaffold with fragments that are similar to the originals; cleavage and assembly of the molecules follow chemical reactions that are defined by a RECAP-like procedure [93]. In this manner, new chemistry is generated by "scaffold hopping" [94]. In principle, the docking program FlexX [58,59], which performs an incremental construction of a ligand within the binding site, could also arrive at new analogues, if many different building blocks are used instead of the original building blocks; no virtual library of millions of potential candidates needs to be constructed, only favourable intermediate solutions and final candidates are generated. The only unsolved problem in this respect is the lack of reliability of the scoring functions [95].

SUMMARY AND CONCLUSIONS

Virtual screening and fragment-based approaches are powerful techniques in the search for new ligands [10,62,96]; promising candidates can be enriched in compound collections and virtual libraries. The integration of protein crystallography, NMR techniques, and virtual screening will "significantly enhance the pace of the discovery process and the quality of compounds selected for further development" [97].

The similarity principle, that similar compounds should exert similar biological activities, has always been a most successful approach in drug research, despite many exceptions to this general concept [98,99]. Chemogenomics is a new term for the dedicated investigation of certain compound classes in target families, such as the G protein-coupled receptors (GPCR), kinases, phosphodiesterases, serine proteases, ion channels, etc. [100-104].

An analogous concept, the "selective optimization of side activities" (SOSA approach), attempts to develop new drugs in the direction of a side-effect of a certain drug [105,106]. Historical examples for the validity of this approach are e.g. antitussive and constipating morphine analogues, diuretic and antidiabetic sulfonamides, and many others [107]; some very recent examples have been reviewed [106,107].

As already mentioned, good ligands are not necessarily good drugs. High-throughput screening of in-house libraries, which originally contained large numbers of reactive, degraded, coloured, fluorescent, and highly lipophilic molecules, and screening of combinatorial libraries of large, lipophilic molecules produced hits that could not be optimized to drug candidates. Awareness of the real problems came only after Lipinski had defined his set of rules [12]. On the other hand, the massive increase of screening failures due to such inappropriate compound collections or libraries turned this awareness of ADME problems into a hype. Prior statistics of 40% failure in clinical investigation (the most expensive phase of drug development), due to ADME problems [108,109] are cited in the literature, again and again; a closer inspection of the data shows that the ADME-related failure can be neglected if anti-infectives are removed from the original sample [6,109]. This is an indication that medicinal chemists have always considered, the importance of ADME properties. Only HTS and early combinatorial chemistry have generated so many problems in this direction. In addition to the Lipinski rules, several *in vitro* and *in silico* techniques are now available for the estimation of ADME properties [110].

With respect to biological testing, Horrobin has raised the question as to whether we are already living in Castalia, the famous virtual land of Hermann Hesse's novel "The Glass Bead Game", where the masters organize and play the most sophisticated, complex and brilliant games - without any context to reality [111]. Sometimes, this is also the case in modelling and drug design [6]. The new tools of drug research are extremely powerful but they will be successful only if the most important factors, some of which have been reviewed here, are considered in the right manner. In the fascinating search for better and safer drugs, the new paradigms of drug discovery have to merge with traditional medicinal chemistry experience [2,112,113].

REFERENCES

[1]   Roberts, R. M. (1989) *Serendipity. Accidental Discoveries in Science.* John Wiley & Sons, New York.

[2]   Sneader, W. (1996) *Drug Prototypes and Their Exploitation.* John Wiley & Sons, Chichester.

[3]   Kubinyi, H. (1999) Chance favors the prepared mind. From serendipity to rational drug design. *J. Recept. Signal Transduct. Res.* **19**:5-39.

[4]   Lahana, R.(1999) How many leads from HTS? *Drug Discov. Today* **4**:47-448.

[5]   Hopkins, A. L., Groom, C. R. (2002) The druggable genome. *Nature Rev. Drug Discov.* **1**:727-730.

[6]   Kubinyi, H. (2003) Drug research: myths, hype and reality. *Nature Rev. Drug Discov.* **2**:665-668.

[7]   Schaus, J. M., Bymaster, F. P. (1998) Dopaminergic approaches to antipsychotic agents. *Ann. Rep. Med. Chem.* **33**:1-10.

[8]   Bymaster, F. P., Calligaro, D. O., Falcone, J. F., Marsh, R. D., Moore, N. A., Tye, N. C., Seeman P., Wong, D. T. (1996) Radioreceptor binding profile of the atypical antipsychotic olanzapine. *Neuropsychopharmacology* **14**:87-96.

[9]   Bymaster, F. P., Nelson, D. L., DeLapp, N. W., Falcone, J. F., Eckols, K., Truex, L. L., Foreman, M. M., Lucaites V. L., Calligaro, D. O. (1999) Antagonism by olanzapine of dopamine D1, serotonin2, muscarinic, histamine H1 and alpha 1-adrenergic receptors *in vitro. Schizophr. Res.* **37**:107-122.

[10]   Böhm, H.-J., Schneider, G. (Eds) (2000) Virtual Screening for Bioactive Molecules. *Methods and Principles in Medicinal Chemistry.* (Mannhold, R., Kubinyi, H., Timmerman, H., Eds), Vol.10. Wiley-VCH, Weinheim.

[11]   a) Rishton, G. M. (1997) Reactive compounds and in vitro false positives in HTS. *Drug Discov. Today* **2**:382-384; b) Rishton, G. M. (2003) Nonleadlikeness and leadlikeness in biochemical screening *Drug Discov. Today* **8**:86-96.

[12]   Lipinski, C. A., Lombardo F., Dominy, B. W., Feeney, P. J. (1997) Experimental and computational approaches to estimate solubility and permeability in drug discovery and development settings. *Adv. Drug Deliv. Rev.* **23**: 3-26.

[13]   MDL Available Compounds Directory, MDL Information Systems Inc., San Leandro, CA, U.S.A.; www.mdl.com.

[14]   Ajay, Walters, W. P., Murcko, M. A. (1998) Can we learn to distinguish between "drug-like" and "nondrug-like" molecules? *J. Med. Chem.* **41**:3314-3324.

[15]   Sadowski, J., Kubinyi, H. (1998) A scoring scheme for discriminating between drugs and nondrugs. *J. Med. Chem.* **41**:3325-3329.

[16]   Walters, W. P., Murcko, M. A. (2002) Prediction of 'drug-likeness'. *Adv. Drug Deliv. Rev.* **54**: 255-271.

[17]   Rarey, M., Dixon, J. S. (1998) Feature trees: a new molecular similarity measure based on tree matching. *J. Comput.-Aided Mol. Design* **12**: 471-490; www.biosolveit.de.

[18]   Rarey, M., Stahl, M. (2001) Similarity searching in large combinatorial chemistry spaces. *J. Comput.-Aided Mol. Design* **15**: 497-520.

[19]   MDL Screening Compounds Directory, MDL Information Systems Inc., San Leandro, CA, U.S.A.; www.mdl.com.

[20]   Güner, O. F.(Ed.) (2000) *Pharmacophore Perception, Development and Use in Drug Design.* International University Line, La Jolla, CA.

[21]   Catalyst, Accelrys Inc., San Diego, CA, U.S.A.; www.accelrys.com.

[22]   Höltje, H.-D., Sippl, W., Rognan, D., Folkers, G. (2003) *Molecular Modelling. Basic Principles and Applications*, 2nd Edn. Wiley-VCH, Weinheim.

[23]   Lommerse, J. P. M., Price, S. L., Taylor, R. (1997) Hydrogen bonding of carbonyl, ether, and ester oxygen atoms with alkanol hydroxyl groups. *Comput. Chem.* **18**:757-774.

[24]   Böhm, H.-J., Brode, S., Hesse, U., Klebe, G. (1996) Oxygen and nitrogen in competitive situations: which is the hydrogen-bond acceptor? *Chem. Eur. J.* **2**:1509-1513.

[25]   Sadowski, J. (2002) A tautomer and protonation pre-processor for virtual screening, Lecture presented at the 224th ACS National Meeting, Boston.

[26]   Nielsen, J. E., Vriend, G. (2001) Optimizing the hydrogen-bond network in Poisson-Boltzmann equation-based pKa calculations. *Proteins* **43**:403-412.

[27]   Nielsen, J. E., McCammon, J. A. (2003) Calculating pKa values in enzyme active sites. *Protein Sci.* **12**:1894-1901.

[28]   Pospisil, P., Ballmer, P., Scapozza. L., Folkers, G. (2003) Tautomerism in computer-aided drug design. *J. Recept. Signal Transduct. Res.* **23**:361-371.

[29]   Trepalin, S. V., Skorenko, A. V., Balakin, K. V., Nasonov, A. F., Lang, S. A., Ivashchenko, A. A., Savchuk, N. P. (2003) Advanced exact structure searching in large databases of chemical compounds. *J. Chem. Inf. Comput. Sci.* **43**:852-860.

[30]   a) Watson, J. D. (1968) *The Double Helix: A Personal Account of the Discovery of the Structure of DNA.* Atheneum Press, New York; b) Watson, J. D., Berry, A. (2003) DNA. *The Secret of Life.* William Heinemann, London; c) www.phy.cam.ac.uk/camphy/dna/dna13_1.htm and ... /dna14_1.htm.

[31]   Brandstetter, H., Grams, F., Glitz, D., Lang, A., Huber, R., Bode, W., Krell, H. W., Engh, R. A. (2001) The 1.8-A crystal structure of a matrix metalloproteinase 8-barbiturate inhibitor complex reveals a previously unobserved mechanism for collagenase substrate recognition. *J. Biol. Chem.* **276**:17405-17412.

[32]   Pospisil, P., Ballmer, P., ETH Zurich, Switzerland; www.pharma.ethz.ch/pc/Agent2/.

[33]   Kubinyi, H., Hamprecht, F. A., Mietzner, T. (1998) Three-dimensional quantitative similarity-activity relationships (3D QSiAR) from SEAL similarity matrices. *J. Med. Chem.* **41**:2553-2564.

[34]   Kearsley, S. K., Smith, G. M. (1990) An alternative method for the alignment of molecular structures: maximizing electrostatic and steric overlap. *Tetrahedron Comput. Methodol.* **3**:615-633.

[35]   Klebe, G., Mietzner, T., Weber, F. (1994) Different approaches toward an automatic structural alignment of drug molecules: applications to sterol mimics, thrombin and thermolysin inhibitors. *J. Comput.-Aided Mol. Design* **8**:751-778.

[36]   Lemmen, C., Lengauer, T. (1997) Time-efficient flexible superposition of medium-sized molecules. *J. Comput.-Aided Mol. Design* **11**:357-368; www.biosolveit.de; www.tripos.com.

[37]   Lemmen, C., Lengauer, T. (2000) Computational methods for the structural alignment of molecules. *J. Comput.-Aided Mol. Design* **14**:215-232; www.biosolveit.de; www.tripos.com.

[38]   Cramer III, R. D., Patterson, D. E., Bunce, J. D. (1988) Comparative molecular field analysis (CoMFA). 1. Effect of shape on binding of steroids to carrier proteins. *J. Am. Chem. Soc.* **110**:5959-5967.

[39]   a) Kubinyi, H. (Ed.)(1993) *3D QSAR in Drug Design*. Theory, Methods and Applications. ESCOM Science Publishers B.V., Leiden; b) Kubinyi, H., Folkers, G., Martin, Y. C. (Eds) (1998) *3D QSAR in Drug Design*. Ligand-Protein Complexes and Molecular Similarity, Vol.2. Kluwer/ESCOM, Dordrecht; also published as *Persp. Drug Discov. Design* 9-11: 1-416; c) Kubinyi, H., Folkers, G., Martin, Y. C. (Eds) (1998) *3D QSAR in Drug Design*. Recent Advances, Vol. 3. Kluwer/ESCOM, Dordrecht; also published as Persp. Drug Discov. Design 12-14: 1-352.

[40]   Klebe, G., Abraham, U., Mietzner, T. (1994) Molecular similarity indices in a comparative analysis (CoMSIA) of drug molecules to correlate and predict their biological activity. *J. Med. Chem.* **37**:4130-4146.

[41]   Sybyl / Unity, Tripos Inc., St. Louis, MO, U.S.A.; www.tripos.com.

[42]   Goodford, P. J. (1984) Drug design by the method of receptor fit. *J. Med. Chem.* **27**:558-564.

[43]   Kubinyi, H. (1998) Structure-based design of enzyme inhibitors and receptor ligands. *Curr. Opin. Drug Discov. Devel.* **1**:4-15.

[44]   Redshaw, S. (1993) Angiotensin-converting enzyme (ACE) inhibitors and the design of Cilazapril. In: *Medicinal Chemistry. The Role of Organic Chemistry in Drug Research.* (Ganellin, C. R., Roberts, S. M. Eds), 2nd Edn, pp. 163-185. Academic Press, London.

[45]   von Itzstein, M., Wu, W. Y., Kok, G. B., Pegg, M. S., Dyason, J. C., Jin, B., Phan, T. V., Smythe, M. L., White, H. F., Oliver, S. W., Colman, P. M., Varghese, J. N., Ryan, D. M., Woods, J. M., Bethell, R. C., Hotham, V. J., Cameron, J. M., Penn, C. R. (1993) Rational design of potent sialidase-based inhibitors of influenza virus replication. *Nature* **363**:418-423.

[46]   Abdel-Magid, A. F., Maryanoff, C. A., Mehrman, S. J. (2001) Synthesis of influenza neuraminidase inhibitors. *Curr. Opin. Drug Discov. Devel.* **4**:776-791.

[47]   Kim, C. U., Lew, W., Williams, M. A., Liu, H., Zhang, L., Swaminathan, S., Bischofberger, N., Chen, M. S., Mendel, D. B., Tai, C. Y., Laver, W. G., Stevens, R. C. (1997) Influenza neuraminidase inhibitors possessing a novel hydrophobic interaction in the enzyme active site: design, synthesis and structural analysis of carbocyclic sialic acid analogues with potent anti-influenza activity. *J. Am. Chem. Soc.* **119**:681-690.

[48]   Veerapandian, P. (Ed.) (1997) *Structure-Based Drug Design.* Marcel Dekker, New York.

[49]   Babine, R. E., Bender, S. L. (1997) Molecular recognition of protein-ligand complexes: applications to drug design. *Chem. Rev.* **97**:1359-1472.

[50]   Gubernator, K., Böhm, H.-J.(Eds) (1998) Structure-Based Ligand Design. *Methods and Principles in Medicinal Chemistry.* (Mannhold, R., Kubinyi, H., Timmerman, H., Eds), Vol. 6. Wiley-VCH, Weinheim.

[51]   Babine R. E., Abdel-Meguid, S. S. (2004) Protein Crystallography in Drug Discovery. *Methods and Principles in Medicinal Chemistry.* (Mannhold, R., Kubinyi, H., Folkers, G., Eds), Vol.20. Wiley-VCH, Weinheim.

[52]   DesJarlais, R. L., Sheridan, R. P., Seibel, G. L., Dixon, J. S., Kuntz, I. D., Venkataraghavan, R.(1988) Using shape complementarity as an initial screen in designing ligands for a receptor binding site of known three-dimensional structure. *J. Med. Chem.* **31**:722-729.

[53]   Böhm, H.-J. (1992) The computer program LUDI: a new method for the de novo design of enzyme inhibitors. *J. Comput.-Aided Mol. Design* **6**:61-78.

[54]   Böhm, H.-J.(1992) LUDI: rule-based automatic design of new substituents for enzyme inhibitor leads. *J. Comput.-Aided Mol. Design* **6**:593-606.

[55]   Böhm, H.-J. (1994) The development of a simple empirical scoring function to estimate the binding constant for a protein-ligand complex of known three-dimensional structure. *J. Comput.-Aided Mol. Design* **8**:243-256.

[56]   Ewing, T. J., Makino, S., Skillman, A. G., Kuntz, I. D. (2001) DOCK 4.0: search strategies for automated molecular docking of flexible molecule databases. *J. Comput.-Aided Mol. Design* **15**:411-428.

[57]   Jones, G., Willett, P., Glen, R. C., Leach, A. R., Taylor R. (1997) Development and validation of a genetic algorithm for flexible docking. *J. Mol. Biol.* **267**:727-748.

[58]   Rarey, M., Kramer, B., Lengauer, T., Klebe, G. (1996) A fast flexible docking method using an incremental construction algorithm. *J. Mol. Biol.* **261**:470-489; www.biosolveit.de; www.tripos.com.

[59]   Lengauer, T., Rarey, M. (1996) Computational methods for biomolecular docking. *Curr. Opin. Struct. Biol.* **6**:402-406.

[60]   Goodsell, D. S., Olson, A. J. (1990) Automated docking of substrates to proteins by simulated annealing. *Proteins* **8**:195-202.

[61]   Morris, G. M., Goodsell, D. S., Halliday, R. S., Huey, R., Hart, W. E., Belew, R. K., Olson, A. J. (1998) Automated docking using a Lamarckian genetic algorithm and and empirical binding free energy function. *J. Comput. Chem.* **19**:1639-1662.

[62] Schneider, G., Böhm, H.-J. (2002) Virtual screening and fast automated docking methods. *Drug Discov. Today* **7**:64-70.

[63] Claussen, H., Buning, C., Rarey, M., Lengauer, T. (2001) FlexE: efficient molecular docking considering protein structure variations. *J. Mol. Biol.* **308**:377-395; www.biosolveit.de; www.tripos.de.

[64] Hindle, S. A., Rarey, M., Buning, C., Lengauer, T. (2002) Flexible docking under pharmacophore type constraints. *J. Comput.-Aided Mol. Design* **16**:129-149; www.biosolveit.de; www.tripos.de.

[65] Hilpert, K., Ackermann, J., Banner, D. W., Gast, A., Gubernator, K., Hadvary, P., Labler, L., Müller, K., Schmid, G., Tschopp, T. B., van de Waterbeemd, H. (1994) Design and synthesis of potent and highly selective thrombin inhibitors. *J. Med. Chem.* **37**:3889-3901.

[66] Boehm, H.-J., Boehringer, M., Bur, D., Gmuender, H., Huber, W., Klaus, W., Kostrewa, D., Kuehne, H., Luebbers, T., Meunier-Keller, N., Mueller, F. (2000) Novel inhibitors of DNA gyrase: 3D structure based biased needle screening, hit validation by biophysical methods, and 3D guided optimization. A promising alternative to random screening. *J. Med. Chem.* **43**:2664-2674.

[67] Green, N. M. (1975) *Avidin. Adv. Protein Chem.* **29**: 85-133.

[68] Hanessian, S., Lu, P.-P., Sanceau, J.-Y., Chemla, P., Gohda, K., Fonne-Pfister, R., Prade, L., Cowan-Jacob, S. W. (1999) An enzyme-bound bisubstrate hybrid inhibitor of adenylosuccinate synthetase. *Angew. Chem. Int. Ed. Engl.* **38**:3159-3162.

[69] Shuker, S. B., Hajduk, P. J., Meadows, R. P., Fesik, S. W. (1996) Discovering high-affinity ligands for proteins: SAR by NMR. *Science* **274**:1531-1534.

[70] Hajduk, P. J., Meadows, R. P., Fesik, S. W. (1997) Discovering high-affinity ligands for proteins. *Science* **278**:497-499.

[71] Hajduk, P. J., Sheppard, G., Nettesheim, D. G., Olejniczak, E. T., Shuker, S. B., Meadows, R. P., Steinman, D. H., Carrera Jr., G. M., Marcotte, P. A., Severin, J., Walter, K., Smith, H., Gubbins, E., Simmer, R., Holzman, T. F., Morgan, D. W., Davidsen, S. K., Summers, J. B., Fesik, S. W. (1997) Discovery of potent nonpeptide inhibitors of stromelysin using SAR by NMR. *J. Am. Chem. Soc.* **119**:5818-5827.

[72] Zerbe, O. (Ed.) (2003) BioNMR in Drug Research. *Methods and Principles in Medicinal Chemistry.* (Mannhold, R., Kubinyi, H., Folkers, G., Eds), Vol. 16. Wiley-VCH, Weinheim.

[73] Fejzo, J., Lepre, C. A., Peng, J. W., Bemis, G. W., Ajay, Murcko, M. A., Moore, J. M. (1999) The SHAPES strategy: an NMR-based approach for lead generation in drug discovery. *Chem. Biol.* **6**:755-769.

[74] Lepre, C. A., Peng, J., Fejzo, J., Abdul-Manan, N., Pocas, J., Jacobs, M., Xie, X., Moore, J. M. (2002) Applications of SHAPES screening in drug discovery. *Comb. Chem. High Throughput Screen* **5**:583-590.

[75] Diercks, T., Coles, M., Kessler, H. (2001) Applications of NMR in drug discovery. *Curr. Opin. Chem. Biol.* **5**:285-291.

[76] Pellecchia, M., Sem, D. S., Wüthrich, K. (2002) NMR in drug discovery. *Nature Rev. Drug Discov.* **1**:211-219.

[77] Jahnke, W., Floersheim, P., Ostermeier, C., Zhang, X., Hemmig, R., Hurth, K., Uzunov, D. P. (2002) NMR reporter screening for the detection of high-affinity ligands. *Angew. Chem. Int. Ed. Engl.* **41**:3420-3423.

[78] Meyer, B., Peters, T. (2003) NMR spectroscopy techniques for screening and identifying ligand binding to protein receptors. *Angew. Chem. Int. Ed. Engl.* **42**:842-890.

[79] Kubinyi, H. (1998) Combinatorial and computational approaches in structure-based drug design. *Curr. Opin. Drug Discovery Dev.* **1**:16-27.

[80] Metz, G., Ottleben, H., Vetter, D. (2003) Small molecule screening on chemical microarrays. In: Protein-Ligand Interactions. From Molecular Recognition to Drug Design, (Böhm, H.-J., Schneider, G., Eds) *Methods and Principles in Medicinal Chemistry.* (Mannhold, R., Kubinyi, H., Folkers, G., Eds), Vol.19, pp.213-236. Wiley-VCH, Weinheim.

[81] Huc, I., Lehn, J.-M. (1997) Virtual combinatorial libraries: dynamic generation of molecular and supramolecular diversity by self-assembly. *Proc. Natl. Acad. Sci. USA* **94**:2106-2110; erratum in *Proc. Natl. Acad. Sci. USA* **94**:8272.

[82] Lehn, J.-M., Eliseev, A. V. (2001) Dynamic combinatorial chemistry. *Science* **291**: 2331-2332.

[83] Ramström, O., Lehn, J.-M. (2002) Drug discovery by dynamic combinatorial libraries. *Nature Rev. Drug Discov.* **1**:26-36.

[84] Hochgürtel, M., Kroth, H., Piecha, D., Hofmann, M. W., Nicolau, C., Krause, S., Schaaf, O., Sonnenmoser, G., Eliseev, A. V. (2002) Target-induced formation of neuraminidase inhibitors from in vitro virtual combinatorial libraries. *Proc. Natl. Acad. Sci. USA* **99**:3382-3387.

[85] Erlanson D. A., Braisted, A. C., Raphael, D. R., Randal, M., Stroud, R. M., Gordon, E. M., Wells, J. A. (2000) Site-directed ligand discovery. *Proc. Natl. Acad. Sci. USA* **97**: 9367-9372.

[86] Erlanson, D. A., Lam, J. W., Wiesmann, C., Luong, T. N., Simmons, R. L., DeLano, W. L., Choong, I. C., Burdett, M. T., Flanagan, W. M., Lee, D., Gordon, E. M., O'Brien, T. (2003) *In situ* assembly of enzyme inhibitors using extended tethering. *Nature Biotechnol.* **21**:308-314.

[87] Maly, D. J., Choong, I. C., Ellman, J. A. (2000) Combinatorial target-guided ligand assembly: identification of potent subtype-selective c-Src inhibitors. *Proc. Natl. Acad. Sci. USA* **97**:2419-2424.

[88] Swayze, E. E., Jefferson, E. A., Sannes-Lowery, K. A., Blyn, L. B., Risen, L. M., Arakawa, S., Osgood, S. A., Hofstadler, S. A., Griffey. R. H. (2002) SAR by MS: a ligand based technique for drug lead discovery against structured RNA targets. *J. Med. Chem.* **45**:3816-3819.

[89]   Lewis, W. G., Green, L. G., Grynszpan, F., Radic, Z., Carlier, P. R., Taylor, P., Finn M. G., Sharpless, K. B. (2002) Click chemistry *in situ*: acetylcholinesterase as a reaction vessel for the selective assembly of a femtomolar inhibitor from an array of building blocks. *Angew. Chem. Int. Ed. Engl.* **41**:1053-1057.

[90]   Wolber, G., Langer, T. (2001) Combigen: a novel software package for the rapid generation of virtual combinatorial libraries. In: *Rational Approaches to Drug Design.* (Höltje, H.-D., Sippl, W., Eds), pp.390-399. (Proceedings of the 13th European Symposium on Quantitative Structure-Activity Relationships, Düsseldorf, 2000. Prous Science, Barcelona.

[91]   Schneider, G., Clement-Chomienne, O., Hilfiger, L., Schneider, P., Kirsch, S., Boehm, H.-J., Neidhart, W. (2000) Virtual screening for bioactive molecules by evolutionary de novo design. *Angew. Chem. Int. Ed. Engl.* **39**:4130-4133.

[92]   Schneider, G., Lee, M. L., Stahl, M., Schneider, P. (2000) De novo design of molecular architectures by evolutionary assembly of drug-derived building blocks. *J. Comput.-Aided Mol. Design* **14**:487-494.

[93]   Lewell, X. Q., Judd, D. B., Watson, S. P., Hann, M. M. (1998) RECAP -retrosynthetic combinatorial analysis procedure: a powerful new technique for identifying privileged molecular fragments with useful applications in combinatorial chemistry. *J. Chem. Inf. Comput. Sci.* **38**:511-522.

[94]   Schneider, G., Neidhart, W., Giller, T., Schmid, G. (1999) "Scaffold-Hopping" by Topological Pharmacophore Search: A Contribution to Virtual Screening. *Angew. Chem. Int. Ed. Engl.* **38**:2894-2896.

[95]   Wang, R., Lu, Y., Wang, S. (2003) Comparative evaluation of 11 scoring functions for molecular docking. *J. Med. Chem.* **46**:2287-2303.

[96]   Bleicher, K. H., Böhm, H.-J., Müller, K., Alanine, A. I. (2003) Hit and lead generation: beyond high-throughput screening. *Nature Rev. Drug Discov.* **2**: 369-378.

[97]   Muchmore, S. W., Hajduk, P. J. (2003) Crystallography, NMR and virtual screening: integrated tools for drug discovery. *Curr. Opin. Drug Discov.Dev.* **6**:544-549.

[98]   Kubinyi, H. (1998) Similarity and dissimilarity: a medicinal chemist's view, In: 3D QSAR in Drug Design. *Ligand-Protein Interactions and Molecular Similarity.* (Kubinyi, H., Folkers G., Martin, Y. C., Eds), Vol. II, pp.225-252. Kluwer/ESCOM, Dordrecht,; also published in *Persp. Drug Design Discov.* 9-11:225-252.

[99]   Martin, Y. C., Kofron, J. L., Traphagen, L. M. (2002) Do structurally similar molecules have similar biological activity? *J. Med. Chem.* **45**:4350-4358.

[100]   Caron, P. R., Mullican, M. D., Mashal, R. D., Wilson, K. P., Su, M. S., Murcko, M. A. (2001) Chemogenomic approaches to drug discovery. *Curr. Opin. Chem. Biol.* **5**:464-470.

[101]   Bleicher, K. H. (2002) Chemogenomics: bridging a drug discovery gap. *Curr. Med. Chem.* **9**:2077-2084.

[102]   Jacoby, E., Schuffenhauer, A., Floersheim, P. (2003) Chemogenomics knowledge-based strategies in drug discovery. *Drug News Perspect.* **16**:93-102.

[103] Müller, G. (2003) Medicinal chemistry of target family-directed masterkeys. *Drug Discov. Today* **8**: 681-691.

[104] Kubinyi, H., Müller, G. (Eds) (2004) Chemogenomics in Drug Discovery - A Medicinal Chemistry Perspective. *Methods and Principles in Medicinal Chemistry.* (Mannhold, R., Kubinyi, H., Folkers, G. Eds), Vol.22. Wiley-VCH, Weinheim.

[105] Wermuth, C. G. (2001) The "SOSA" aproach: an alternative to high-throughput screening. *Med. Chem. Res.* **10**:431-439.

[106] Wermuth, C. G. (2004) Selective optimization of side activities: another way for drug discovery. *J. Med. Chem.* **47**:1303-1314.

[107] Kubinyi, H. (2004) Drug discovery from side effects. In: *Chemogenomics in Drug Discovery - A Medicinal Chemistry Perspective* (Kubinyi, H., Müller, G., Eds)f Methods and Principles in Medicinal Chemistry. (Mannhold, R., Kubinyi, H., Folkers, G., Eds), Vol.22, pp.43-67. Wiley-VCH, Weinheim.

[108] Prentis, R. A., Lis, Y., Walker, S. R. (1988) Pharmaceutical innovation by the seven UK-owned pharmaceutical companies (1964-1985). *Br. J. Clin. Pharmac.* **25**:387-396.

[109] Kennedy, T. (1997) Managing the drug discovery/development interface. *Drug Discov. Today* **2**:436-444.

[110] van de Waterbeemd, H., Gifford, E. (2003) ADMET in silico modelling: towards prediction paradise? *Nature Rev. Drug Discov.* **2**:192-204.

[111] Horrobin, D. F. (2003) Modern biomedical research: an internally self-consistent universe with little contact with medical reality? *Nature Rev. Drug Discov.* **2**:151-154.

[112] Ryan, J. F. (Ed.) (2000) *The Pharmaceutical Century. Ten Decades of Drug Discovery.* Supplement to ACS Publications, American Chemical Society, Washington, DC.

[113] Wermuth, C. G. (Ed.) (2003) *The Practice of Medicinal Chemistry*, 2nd Edn. Academic Press, London.

The Chemical Theatre of Biological Systems, May 24th - 28th, 2004, Bozen, Italy

THERMODYNAMICS IN DRUG DESIGN.
HIGH AFFINITY AND SELECTIVITY

ERNESTO FREIRE

Johns Hopkins University, Department of Biology, Baltimore, MD 21218, U.S.A.

E-Mail: ef@jhu.edu

Received: 27th July 2005 / Published: 22nd July 2005

ABSTRACT

In drug discovery, active compounds identified by screening or other approaches usually bind to their targets with micromolar or weaker affinities. To become effective drugs, the binding affinities of those compounds need to be improved by three or more orders of magnitude. This task is not trivial if one considers that it needs to be done while satisfying several stringent constraints related to bioavailability, membrane permeability, water solubility, pharmacokinetics, toxicity, etc.. In addition, successful candidates need to exhibit appropriate selectivity and in the case of anti-infectives low susceptibility to mutations associated with drug resistance. These constraints emphasize the need for accurate ways of predicting the various effects of introducing diverse chemical functionalities or scaffold modifications during lead optimization, in particular effects on affinity and selectivity. Recently, it has become evident that the attainment of extremely high affinity, selectivity or adaptability is related to the proportion in which the enthalpy and entropy changes contribute to the binding affinity, and that appropriate control over these variables is critical during the design process. Since modern microcalorimetry provides extremely accurate measurements of the enthalpy and entropy contributions to binding affinity, it provides the basis for the development of rigorous algorithms aimed at: 1) Binding affinity optimization; 2) Improvement of binding selectivity between similar targets; 3) Incorporation of binding adaptability to mutations that cause drug resistance. In this chapter, the role of thermodynamics and enthalpy/entropy profiling in lead optimization will be discussed.

THE BINDING AFFINITY

The identification of drug candidates by screening large libraries of compounds and their subsequent optimization is an important step in the development of new pharmaceutical drugs. Modern high-throughput screening procedures are able to evaluate millions of compounds and identify those that exhibit the highest affinity or inhibitory potency with relative accuracy. Usually, compounds identified by screening have only marginal potency and their binding affinity needs to be improved by orders of magnitude. From a thermodynamic point of view, the binding affinity, K_a, is a function of the Gibbs energy of binding:

$$K_a = e^{-\Delta G / RT} \tag{1}$$

where R is the gas constant and T the absolute temperature. The Gibbs energy of binding is in turn defined by the enthalpy (ΔH) and entropy (ΔS) changes:

$$\Delta G = \Delta H \angle T\Delta S \tag{2}$$

therefore,

$$K_a = e^{-(\Delta H - T\Delta S) / RT} \tag{3}$$

$$K_a = e^{-\Delta H / RT} \times e^{\Delta S / R}$$

Since initial leads typically have affinities in the micromolar range and effective drugs require nanomolar and sometimes even higher affinities, an increase of at least three orders of magnitude in binding affinity is required. This increase is equivalent to an additional 4 kcal/mol in the Gibbs energy of binding. According to Equation 2, the binding affinity can be optimized by making ΔH more negative, ΔS more positive or by a combination of both. However, since enthalpy and entropy changes originate from different interactions, enthalpically or entropically optimized compounds are not equivalent even if they have the same affinity against the intended target.

IMPORTANT CONTRIBUTIONS TO BINDING ENTHALPY AND ENTROPY

Enthalpic and entropic contributions to the Gibbs energy originate from different types of interactions. Enthalpic contributions primarily reflect the strength of the inhibitor interactions with the protein (hydrogen bonds, van der Waals interactions) relative to those with the solvent (in this paper our discussion refers to the intrinsic binding enthalpy and not to enthalpic effects associated with protein conformational changes or linked protonation events, see reference [1] for a complete discussion). Because the unfavourable enthalpy change associated with the desolvation of polar groups is very large ($\sim$ 8 - 9 kcal/mol at 25°C for NH_2, NH, OH, etc. [2], the interactions of polar groups with a protein target need to be strong enough to overcome the unfavourable desolvation enthalpy of those groups. For that reason the enthalpic optimization of a ligand cannot be accomplished by simply increasing the number of polar functionalities in the compound. Unless those functionalities (e.g. hydroxyl or amino groups) establish strong hydrogen bonds with the target, the desolvation penalty will be the predominant term and the overall enthalpy will be unfavourable. The compensation between enthalpy of interaction and enthalpy of desolvation can be appreciated by examining the binding energetics of existing HIV-1 protease inhibitors [3]. Saquinavir for example binds to the protease with an unfavourable enthalpy change of 1.9 kcal/mol at 25°C [4]. Under the same conditions, TMC-126 binds to the protease with a favourable binding enthalpy of -12 kcal/mol [4]. Saquinavir has a total of 11 hydrogen bond donors and acceptors while TMC-126 has 10. It is evident that the binding enthalpy does not correlate with the number of groups that can participate in hydrogen bonds, emphasizing that the introduction of hydrogen bond donors or acceptors at arbitrary or even weak hydrogen bonding positions can be detrimental to the binding enthalpy and binding affinity. In fact saquinavir binds to the protease with a dissociation constant, K_d, of 0.4 nM whereas TMC-126 binds to the protease with a K_d of 0.004 nM [4].

Two major terms contribute to the binding entropy. The most important one is the change in solvent entropy arising from the complete or partial desolvation of the drug molecule and some regions of the protein upon binding. The second contribution is due to changes in conformational degrees of freedom experienced by the drug molecule and protein upon binding. The change in solvation entropy is tremendously favourable if the surfaces that are buried upon binding are predominantly hydrophobic.

Moreover, the unfavourable enthalpy associated with the desolvation of hydrophobic groups (e.g. CH_3, CH_2, CH) is one order of magnitude smaller than that of polar groups. The combination of a large favourable entropy and a small desolvation enthalpy contributes to yield a highly favourable Gibbs energy for hydrophobic burial which is estimated to be around 25 cal/mol×Å^2 [5-7]. The major drawback of this approach to affinity optimization is that the compound rapidly becomes insoluble in water.

Changes in conformational degrees of freedom occur in both the drug molecule and the protein molecule. In solution, the drug molecule is capable of adopting multiple conformations due to rotations around bonds. Upon binding, interactions with the protein constrain the drug to a single conformation resulting in a loss of conformational entropy. With the protein, a similar loss in conformational degrees of freedom is usually observed as side chains within the binding cavity and some backbone elements (e.g. flexible loops that become immobilized) lose conformational degrees of freedom. Since any loss in conformational entropy is unfavourable, one approach to minimize those adverse effects has been to introduce conformational constraints in the drug molecule. One rotatable bond that becomes immobilized upon binding carries a Gibbs energy penalty close to 0.5 kcal/mol due to the loss of conformational entropy [8]. Everything else being equal, a conformationally constrained molecule has a higher binding affinity because it does not carry that entropy penalty.

It must be noticed that the binding of most compounds is associated with a change in heat capacity (ΔC_p) which makes the enthalpy and entropy changes temperature dependent. Since ΔC_p is usually negative, the enthalpy change will become more negative at higher temperatures and in some cases a change in sign can be observed. For example, the binding enthalpy of saquinavir will change from 1.9 kcal/mol at 25°C to -2.2 kcal/mol at 37°C due to a binding ΔC_p of -340 cal/K×mol [9]. The change in the numerical values of the thermodynamic parameters with temperature does not imply that the mode of binding of a particular inhibitor has changed. Structural or molecular correlations of thermodynamic parameters should be made at one particular temperature, usually 25°C, at which characteristic values for different interactions have been tabulated and are well known [2]. In this paper we have followed that convention and use 25°C as the reference temperature for analysis.

Additional contributions to the binding enthalpy or entropy like those associated with protein conformational changes coupled to binding [1] need to be considered explicitly if the goal is to account for absolute values of the Gibbs energy, enthalpy or entropy changes. Those terms are less critical in scoring functions since drug molecules targeting the same protein site will generally elicit the same conformational change in the protein and contribute a constant term [1]. The same situation occurs with contributions arising from the reduction in rotational/translational degrees of freedom which are similar for all ligands. Other interactions like salt bridges are less common in drug design due to the infrequent use of charged molecules as drugs, but need to be considered if present.

THE LOCK AND KEY PARADIGM

The most important drug design paradigm currently in use is derived from the classic key-and-lock hypothesis of enzyme specificity originally advanced by Emil Fischer in 1890 (see [10] for a review). The design paradigm is commonly referred to as the "shape complementarity principle" and essentially entails the synthesis of conformationally constrained drug molecules pre-shaped to the geometry of the target-binding site. A molecule that is pre-shaped to the target and conformationally constrained provides specificity and simultaneously enhances affinity. It provides specificity because the probability of finding different proteins with geometrically identical binding sites is generally low. There are important exceptions such as the kinases, all of which bind ATP and have similar binding sites, or blood coagulation factors, all of which are serine proteases with highly homologous catalytic sites. Drug design against individual members of these protein classes is notoriously difficult due to selectivity issues, emphasizing the need for better ways of approaching and solving the issue of selectivity. In addition to improving selectivity, conformational constraints also improve affinity because they minimize the loss of conformational degrees of freedom of the drug molecule upon binding. Organic and medicinal chemists have been able to successfully implement this strategy and design conformationally constrained molecules against a variety of targets.

Shape complementarity, however, does not guarantee binding. For binding to occur a favourable Gibbs energy is required.

Since the effective binding energetics is the difference between the magnitude of the drug/target interactions and the interactions with the solvent, it is always possible to generate a significant binding affinity by making the interactions of the drug with the solvent unfavourable; i.e. by increasing the hydrophobicity of the drug. This is in fact a common strategy, and as a result a high proportion of affinity-optimized drug candidates is highly hydrophobic and rigid (pre-shaped to the geometry of the binding site). The tendency towards higher hydrophobicity in new drug candidates is well known [11-13].

EXTREMELY HIGH AFFINITY

Compounds optimized according to the strategy described above have characteristic thermodynamic signatures as illustrated for compound 1 in Fig. 1.

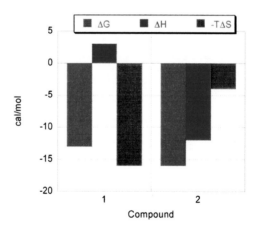

Figure 1. Thermodynamic signatures for two different hypothetical compounds. Compound 1 is entropically optimized, its binding is characterized by an extremely large entropy value that needs to compensate the unfavourable binding enthalpy. Compound 2 is favoured both enthalpically and entropically. As such, it can achieve extremely high affinity without the enthalpy or entropy changes assuming extreme values.

The combination of conformational constraints and hydrophobicity results in compounds that are entropically optimized and characterized by unfavourable or only slightly favourable enthalpy changes. All first generation HIV-1 protease inhibitors (nelfinavir, saquinavir, indinavir and ritonavir) fall under this category [4,9]. The binding of these compounds is entropically driven and at 25°C their binding enthalpy is either unfavourable (indinavir, saquinavir, nelfinavir) or only slightly favourable [4,9].

There are several unwanted consequences associated with thermodynamic signatures like the one associated with compound 1 in Fig. 1. First, the enthalpic and entropic contributions to the binding affinity point in opposite directions, partially compensating each other. As a result, the binding affinity cannot achieve extremely high values. A significant portion of the favourable entropy change is used to offset the unfavourable enthalpy without a real gain in binding affinity. Since hydrophobicity is the main contributor to the favourable entropy change, these drugs exhibit the characteristic problems associated with poor water solubility. Finally, in the case of anti-viral drugs and other anti-infectives the lack of flexibility associated with the presence of excessive conformational constraints in the drug molecule prevents accommodation to binding site variations due to mutations and consequently a high susceptibility to drug resistance [14,15].

The thermodynamic signature of compound 2 in Fig. 1 is different. In this case, both the enthalpy and entropy changes contribute favourably to the binding affinity. As a result neither contribution has to assume extreme values in order to achieve a much larger binding affinity. In fact, the combination of favourable binding enthalpy and entropy is the key to reach extremely high binding affinities. In particular, the entropy change does not need to be exceptionally large to maintain high affinity allowing the designer the possibility of reducing the hydrophobicity of the compound and improve aqueous solubility. It provides the designer with a wider spectrum of possibilities since enthalpically favourable compounds do not need to be more polar as demonstrated for the case of HIV-1 protease inhibitors. In the case of anti-infectives, a favourable binding enthalpy also permits the introduction of flexible elements that will lower the susceptibility to mutations associated with drug resistance [16].

SELECTIVITY

Not all the forces that contribute to binding affinity contribute in the same proportion to specificity or selectivity. The two most relevant examples are hydrogen bonding and hydrophobicity, major contributors to the binding enthalpy and entropy respectively. The strength of hydrogen bonds and therefore their contribution to binding affinity is highly dependent on the distance and angle between donor and acceptor groups. As such, the engineering of several hydrogen bonds at critical locations will define a stereochemical fingerprint that will contribute significantly to affinity as well as selectivity.

On the other hand, the hydrophobic effect is non-specific and driven by the tendency of the compound to escape water rather than by a specific attraction to the target molecule. In fact, within the classic design paradigm, hydrophobicity provides the binding force while "shape complementarity" through conformational constraints provides specificity and selectivity.

From the above considerations, it is evident that maximal selectivity or specificity will be achieved if shape complementarity is combined with a cluster of strong stereospecific hydrogen bonds between drug and protein. Analysis performed in this laboratory and implemented in in-house design algorithms indicate that maximal selectivity is achieved with a critical number of hydrogen bonds. Either a very small or a large number of hydrogen bond donors or acceptors in the drug molecule can lead to poor selectivity. For example, if a molecule has a significant number of hydrogen bond donors and acceptors but binds to its target with only weak or unfavourable binding enthalpy, it indicates that some of those groups only make weak hydrogen bonds and are unable to compensate their unfavourable desolvation enthalpy. Against a homologous target, the same drug molecule may establish a different combination of similarly weak hydrogen bonds, giving rise to a similar energetic situation and resulting in poor selectivity. If, on the other hand, a drug molecule has few hydrogen bond donors and acceptors but binds with a strong favourable enthalpy, it would be indicative of the formation of strong hydrogen bonds with the target. Since the formation of these strong bonds requires a precise geometric arrangement for proper stereochemistry, the likelihood of finding a similar arrangement in other protein is limited, resulting in higher selectivity. In addition, since none of the groups participate in weak bonds, the possibility of alternative patterns is also minimized.

The above situation is illustrated schematically in Fig. 2. In this figure, drug molecule 1 with six hydrogen-bond capable groups will bind to proteins A and B with similar weak binding enthalpy (in both cases only three bonds are satisfied and three groups do not form bonds but pay desolvation penalty). Drug molecule 2, on the other hand, with only three hydrogen-bond capable groups will bind protein A with strong favourable enthalpy (the three bonds are satisfied) but no protein B (only one bond is satisfied and two groups only pay desolvation penalty). This example illustrates the advantages of fewer, geometrically well defined, strong bonds to improve selectivity. Of course, the cluster of bonds should define a pattern unique to the intended target.

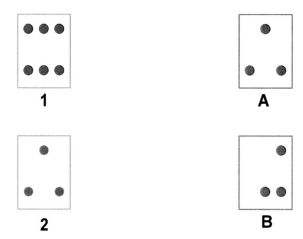

Figure 2. Selectivity is better achieved when a compound has few hydrogen bond donors and acceptors, all of which form strong bonds with the target. In this example, compound 1 has six groups and will bind with weak or unfavourable enthalpy to both protein A and protein B on the right (in both cases three of the groups do not form bonds and contribute unfavourably to the binding enthalpy due to the desolvation penalty). Compound 2 has three groups that perfectly match protein A but not protein B. Compound 1 binds with highly favourable enthalpy to protein A but will display unfavourable enthalpy towards protein B.

BINDING THERMODYNAMICS AND THE RULE OF 5

The Lipinski's "rule of five" [11-13] provides a powerful set of criteria for the solubility and permeability and, consequently, oral bioavailability of drug candidates. It basically stipulates that poor absorption or permeability is more likely when: 1) there are more than 5 hydrogen bond donors (expressed as the sum of NH's and OH's); 2) the molecular weight is over 500; 3) the log P is over 5 (or MlogP is over 4.15); and, 4) there are more than 10 hydrogen bond acceptors (expressed as the sum of N's and O's) [11].

Since some of the terms in the rule of five are related to hydrophobicity or groups that participate in hydrogen bonding interactions, it is important to assess if there is a correlation between thermodynamic parameters for binding and the rule of five. The four panels in Fig. 3 show the dependence of the binding enthalpy of HIV-1 protease inhibitors with the four criteria that define the rule of five. It is clear in the figure that there is no correlation between the number of hydrogen bond donors and the binding enthalpy. In fact, close inspection of the figure reveals that with four donor groups, for example, a compound can be characterized by a large positive (unfavourable) or large negative (favourable) binding enthalpy.

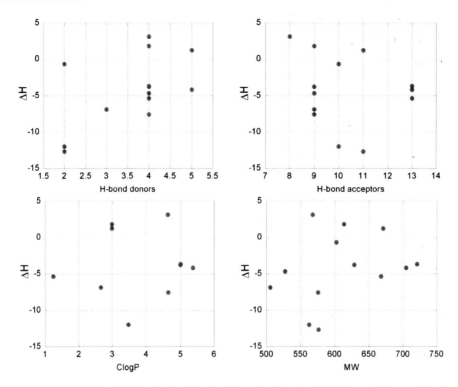

Figure 3. The correlation between the binding enthalpy (cal/mol) of HIV-1 protease inhibitors (included in the graph are indinavir, saquinavir, nelfinavir, ritonavir, amprenavir, lopinavir, atazanavir, KNI-577, KNI-272, KNI-764, TMC-114, TMV-126) with the number of hydrogen bond donors, hydrogen bond acceptors, octanol/water partition coefficient (C logP) and molecular weight, the four variables that define Lipinski's rule of five [11-13].

A similar situation can be observed with the number of hydrogen bond acceptors. As discussed above, the binding enthalpy is the sum of the favourable enthalpy of interaction and the unfavourable enthalpy of desolvation. If the hydrogen bonds are not strong or if a potential group is buried without making a bond, the unfavourable enthalpy of desolvation will be the dominant term. The figure also indicates the lack of correlation of the binding enthalpy with the calculated octanol/water partition coefficient and with the molecular weight of the inhibitors. Together, these observations indicate that a compound can be enthalpically or entropically optimized and that this character is not correlated with the parameters that define the rule of five. This conclusion is particularly important for enthalpically optimized compounds, due to the erroneous belief that these compounds require a large number of hydrogen bond donors and acceptors.

As discussed above, enthalpically optimized compounds afford significant advantages in affinity and selectivity, and in the case of anti-infectives also in adaptability to mutations associated with drug resistance as discussed elsewhere [3].

ISOTHERMAL TITRATION CALORIMETRY

The discussion above emphasizes the advantages of enthalpic optimization and, consequently the need for accurate measurements of binding enthalpies and binding entropies in addition to binding affinity. During the last decade, the development of highly sensitive isothermal titration calorimeters has revolutionized the analysis of binding reactions. Unlike any other technique, isothermal titration calorimetry (ITC) provides a complete thermodynamic characterization of the binding reaction since it directly measures the enthalpy of binding and the association constant, and consequently the entropy change. Furthermore, experiments performed at different temperatures provide the temperature dependence of the enthalpy change, i.e. the heat capacity change associated with the binding reaction. Unlike spectroscopic or other techniques, there is no need to obtain the temperature dependence of the association constant in order to estimate enthalpy changes through the van't Hoff equation. The van't Hoff analysis lacks accuracy due to three facts: 1) the narrow temperature range permissible in biological experiments; 2) the association constant very often appears temperature-independent within the accessible experimental range due to effects related to the existence of a change in heat capacity with binding; and 3) poor precision in the association constant determination that translate into large errors for ΔH and ΔS. Isothermal titration calorimetry does not require a van't Hoff analysis since it measures directly the heat released or absorbed by a binding reaction and therefore the enthalpy change. As a result, enthalpy changes can be routinely measured with accuracies close to 0.1 kcal/mol [17].

CONCLUSIONS

Even though many combinations of ΔH and ΔS values will elicit the same binding affinity (i.e. the same ΔG and therefore the same K_a), the properties and the response of these compounds to changes in the environment or in the protein target are not the same because they originate from different types of interactions.

Enthalpically optimized compounds have important advantages over their entropically optimized counterparts. First, because their affinity does not rely on hydrophobicity much higher affinities can be achieved without completely losing water solubility. Second, because enthalpic interactions arise from stereochemically specific interactions (e.g. hydrogen bonds), it is possible to achieve higher selectivity, provided that the compound contains only a critical number and arrangement of hydrogen bond donors and acceptors.

ACKNOWLEDGEMENTS

This work was supported by grants from the National Institutes of Health GM57144, GM56550 and the National Science Foundation MCB-0131241.

REFERENCES

[1]   Luque, I., Freire, E. (2002) Structural parameterization of the binding enthalpy of small ligands. *Proteins* **49**:181-190.

[2]   Cabani, S., Gianni, P., Mollica, V., Lepori, L. (1981) Group contributions to the thermodynamic properties of non-ionic organic solutes in dilute aqueous solution. *J. Solution Chem.* **10**:563-595.

[3]   Ohtaka, H., Muzammil, S., Schon, A., Velazquez-Campoy, A., Vega, S., Freire, E. (2004) Thermodynamic rules for the design of high affinity HIV-1 protease inhibitors with adaptability to mutations and high selectivity towards unwanted targets. *Int. J. Biochem. Cell Biol.* **36**:1787-1799.

[4]   Velazquez-Campoy, A., Muzammil, S., Ohtaka, H., Schon, A., Vega, S., Freire, E. (2003) Structural and thermodynamic basis of resistance to HIV-1 protease inhibition: implications for inhibitor design. *Curr. Drug Targets - Infect. Disord.* **3**: 311-328.

[5]   Chothia, C. (1976) The nature of the accessible and buried surfaces in proteins. *J. Mol. Biol.* **105**:1-14.

[6]   Eisenberg, D., McLachlan, A.D. (1986) Solvation energy in protein folding and binding. *Nature* **319**:199-203.

[7]   Raschke, T.M., Tsai, J., Levitt, M. (2001) Quantification of the hydrophobic interaction by simulations of the aggregation of small hydrophobic solutes in water. *Proc. Natl. Acad. Sci. USA* **98**:5965-5969.

[8]   Daquino, J.A., Freire, E., Amzel, L.M. (2000) Binding of small molecules to macromolecular targets: evaluation of conformational entropy changes. *Proteins* **4**:93-107.

[9]   Todd, M.J., Luque, I., Velazquez-Campoy, A., Freire, E. (2000) The thermodynamic basis of resistance to HIV-1 protease inhibition. Calorimetric analysis of the V82F/ I84V active site resistant mutant. *Biochemistry* **39**:11876-11883.

[10] Koshland, D.E., Jr. (1994) The key-lock theory and the induced fit theory. *Angew. Chem. Int. Ed. Engl.* **33**:2375-2378.

[11] Lipinski, C.A., Lombardo, F., Dominy, B.W., Feeney, P.J. (1997) Experimental and computational approaches to estimate solubility and permeability in drug discovery and development settings. *Adv. Drug Deliv. Rev.* **23**:3-25.

[12] Lipinski, C.A. (2000) Drug-like properties and the causes of poor solubility and poor permeability. *J. Pharmacol. Toxicol. Methods* **44**:235-249.

[13] Lipinski, C.A. (2003) Physicochemical properties and the discovery of orally active drugs: technical and people issues. In: *Molecular Informatics Confronting Complexity. Proceedings of the Beilstein International Workshop* May 13th - May 16th, 2002, Bozen, Italy. (Hicks, M.G., Kettner, C., Eds), pp. 59-78. Logos Verlag, Berlin.

[14] Velazquez-Campoy, A., Kiso, Y., Freire, E. (2001) The binding energetics of first and second generation HIV-1 protease inhibitors: implications for drug design. *Archs Biochim. Biophys.* **390**:169-175.

[15] Ohtaka, H., Velazquez-Campoy, A., Xie, D., Freire, E. (2002) Overcoming drug resistance in HIV-1 chemotherapy: the binding thermodynamics of Amprenavir and TMC-126 to wild type and drug-resistant mutants of the HIV-1 protease. *Protein Sci.* **11**:1908-1916.

[16] Vega, S., Kang, L. W., Velazquez-Campoy, A., Kiso, Y., Amzel, L. M., Freire, E. (2004) A structural and thermodynamic escape mechanism from a drug resistant mutation of the HIV-1 protease. *Proteins* **55**:594-602.

[17] Leavitt, S., Freire, E. (2001) Direct measurement of protein binding energetics by isothermal titration calorimetry. *Curr. Opin. Struct. Biol.* **11**:560-566.

⊙ Beilstein-Institut   The Chemical Theatre of Biological Systems, May 24th - 28th, 2004, Bozen, Italy

ENGINEERING THE BIOSYNTHESIS OF NONRIBOSOMAL LIPOPEPTIDE ANTIBIOTICS

JASON MICKLEFIELD[*1] AND COLIN P. SMITH[2]

[1]Department of Chemistry, University of Manchester Institute of Science and Technology (UMIST), Sackville Street, PO Box 88, Manchester M60 1QD, U.K.

[2]School of Biomedical and Molecular Sciences, University of Surrey, Guildford, Surrey GU2 7XH, U.K.

E-Mail: *jason.micklefield@umist.ac.uk

Received: 16th August 2004 / Published: 22nd July 2005

ABSTRACT

Biosynthetic engineering entails reprogramming the genes involved in the biosynthesis of natural products so as to generate new molecules, which would otherwise not exist in nature. Potentially this approach can be used to providing large numbers of secondary metabolites variants, with improved biological activities, many of which are too complex for effective total synthesis.

The calcium dependent antibiotic (CDA), from *Streptomyces coelicolor*, is nonribosomal lipopeptide. CDA is structurally related to the therapeutically important antibiotic daptomycin. The CDA producer, *S. coelicolor*, is also highly amenable to genetic modification, which makes CDA an ideal template for biosynthetic engineering. To this end we have probed the biosynthetic origins of CDA and utilized this information to develop methods which have enabled the first engineered biosynthesis of novel CDA-type lipopeptides. Notably a mutasynthesis approach was developed to generate CDAs with modified arylglycine residues. Active site modification of adenylation domains within the CDA nonribosomal peptide synthases also led to new lipopeptides.

INTRODUCTION

Nonribosomal peptides are secondary metabolites that are biosynthesized independently of the ribosome in prokaryotes and lower eukaryotes [1-3]. These complex natural products are linear or cyclic peptides comprised of many unusual as well as proteinogenic amino acids. Often the peptides contain fatty acid, polyketide, or carbohydrate building blocks. Oxidative cross-linking of amino acid side chains can further increase the diverse array of skeletal structures that are thus formed (Fig. 1). Indeed it transpires that the nonribosomal peptides are among the most structurally diverse and widespread secondary metabolites in nature [1-3]. Not surprisingly they display a wide range of biological activities and include many important therapeutic agents such as the immunosuppressive agent cyclosporin, the antitumour agent bleomycin and the antibiotic vancomycin (Fig. 1).

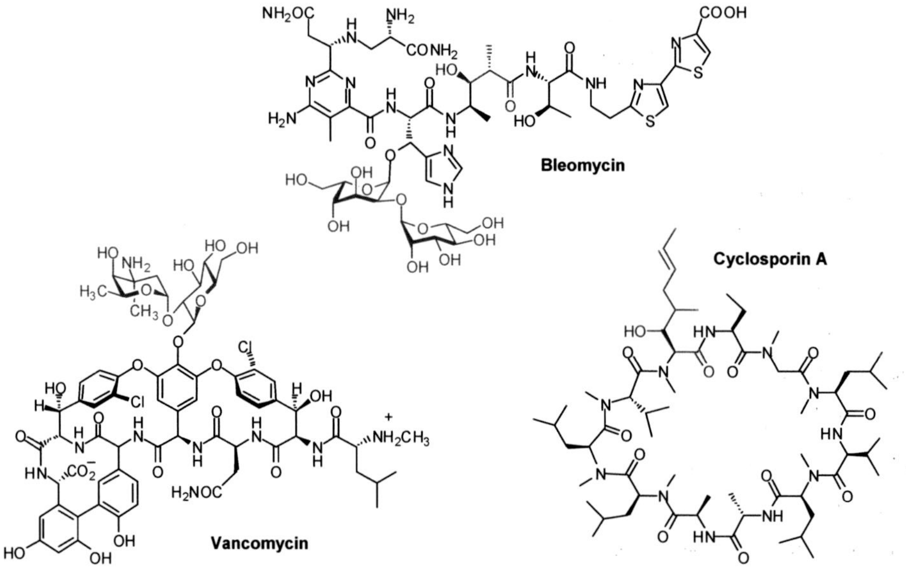

Figure 1. Nonribosomal peptide natural products of therapeutic importance. Bleomycin is produced by *Streptomyces verticillus*, vancomycin comes from *Streptomyces orientalis* and cylcosporin A is derived from *Tolypocladium niveum* and other fungi.

Whilst their structural complexity gives rise to their exquisite biological activities it also means that total synthesis is unable to supply the quantity or diversity of products required for drug development programmes. Consequently we are dependent on nature for our supply of these molecules and if we want novel structural variants of these existing natural products, with improved properties, then alternative biochemical methods of production must be developed.

To this end considerable attention has been focused on understanding how nature assembles the nonribosomal peptides, so that this insight might be used to develop methods which will lead to the engineered, possibly even combinatorial, biosynthesis of novel nonribosomal peptides with improved biological activities [4-7].

The nonribosomal peptide synthetases (NRPS) are the key assembly-line enzymes in the biosynthesis of the nonribosomal peptides [1-3]. NRPS are extremely large enzymes that contain multiple modules each of which is responsible for the activation and incorporation of a specific amino acid into the nascent peptide chain. A minimal NRPS module comprises a condensation (C), an adenylation (A) and a thiolation (T) domain, which is sometimes called the peptidyl carrier protein (PCP) domain. The T-domain is first post-translationally modified by a phosphopantetheinyl transferase (PPtase) enzyme which transfers the phosphopantetheine (Ppant) side chain of co-enzyme A to an active site serine residue. The A-domain then activates a specific amino acid substrate through the formation of amino-acyl adenylate intermediate, which is attacked by the thiol terminus of the Ppant group on the T-domain. It is upon this flexible Ppant side chain that the substrate amino-acyl thioesters can translocate to the upstream C-domain, which mediates the peptide bond forming reaction with the upstream intermediate. Release of the peptide from the NRPS, which is often accompanied by cyclization, is catalysed by a C-terminal thioesterase (Te) domain. Structural biology and enzymology studies have provided key insight into the specificities and mechanisms of these and other catalytic domains within NRPS modules [1-3]. From this a detailed picture has emerged which has made possible the engineered biosynthesis of new products [4-7]. Notably domains within NRPS modules can be surgically replaced or whole modules deleted by manipulation of the NRPS encoding genes, resulting in the mutant organisms that are capable of producing engineered products with different amino acid sequence and ring size [8, 9]. More recently, site directed mutagenesis of A-domains has been employed to change the specificity of the amino acids activated by the NRPS, which has also led to new peptide products [10].

THE ACIDIC LIPOPEPTIDE FAMILY OF ANTIBIOTICS

The calcium-dependent antibiotic (CDA), from *Streptomyces coelicolor* A3(2), is a cyclic-lactone undecapeptide which, in addition to an *N*-terminal 2,3-epoxyhexanoyl side chain, contains several D-configured as well as non-proteinogenic amino acids including D-4-hydroxyphenylglycine (D-HPG), D-3-phosphohydroxyasparagine and L-3-methylglutamic acid [11, 12] (Fig. 2). CDA shares a similar structure, and possibly a related mode of action [13], to other nonribosomally biosynthesized acidic lipopeptide antibiotics including daptomycin from *Streptomyces roseosporus* [14], A54145 from *Streptomyces fradiae* [15] and the friulimicins along with amphomycins from *Actinoplanes friuliensis* [16].

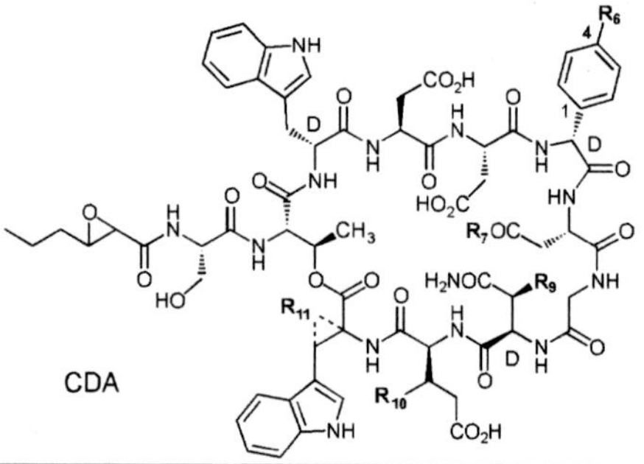

CDAx	R_6	R_7	R_9	R_{10}	R_{11}	Mol. wt.
CDA1b	OH	OH	OPO_3H_2	H	H,H	1562
CDA2a	OH	OH	OPO_3H_2	CH_3	π-bond	1574
CDA2b	OH	OH	OPO_3H_2	CH_3	H,H	1576
CDA3a	OH	OH	OH	H	π-bond	1480
CDA3b	OH	OH	OH	H	H,H	1482
CDA4a	OH	OH	OH	CH_3	π-bond	1494
CDA4b	OH	OH	OH	CH_3	H,H	1496
CDA2d	H	OH	OPO_3H_2	CH_3	H,H	1560
CDA2fa	F	OH	OPO_3H_2	CH_3	π-bond	1576
CDA2fb	F	OH	OPO_3H_2	CH_3	H,H	1578
CDA2a-7N	OH	NH_2	OPO_3H_2	CH_3	π-bond	1573

Figure 2. The structures of the calcium-dependent antibiotics (CDA). CDA1b, 2b, 3b and 4b were isolated and characterized previously [12]. CDA2a and CDA4a were isolated from S. coelicolor strain 2377 grown on solid media. CDA3a was isolated previously, but not fully characterized. CDA2d, CDA2fa and CDA2fb were generated by mutasynthesis [11]. CDA2a-7N is derived from active site modification of the module 7 Asp activating A-domain [35].

All of these lipopeptides contain common amino acid residues including several acidic residues which are important for calcium binding and subsequent antibiotic activity [13]. Remarkably, daptomycin recently became the first new class of natural antibiotic to reach the clinic in many years and is currently being used, under the trade name Cubicin, to treat skin infections with trials underway for its use to treat more serious life-threatening infections [17,18]. Thus there is a real need to develop methods that will enable the reprogrammed, engineered biosynthesis of new lipopeptide of the CDA/daptomycin class with altered and possibly improved antimicrobial activities.

Attempts to engineer new lipopeptides were made possible when the CDA biosynthetic gene cluster was identified, cloned and then sequenced [19,20]. This revealed open reading frames encoding 3 nonribosomal peptide synthetases (cdaPS1, 2 & 3) which are responsible for assembling 6, 3 and 2 amino acids respectively [11] (Fig. 3). Notably, there is an unusual *N*-terminal *C′*-domain that is likely to be involved in the transfer of the fatty acid to the first amino acid Ser of module 1. Also modules 3, 6 and 9 possess additional epimerization (E) domains which are responsible for the D-configured amino acids found at those positions in CDA. Finally there is a *C*-terminal Te-domain which catalyses the release of the peptide from the cdaPS3, through cyclization of the Ser-1 hydroxyl group on to the *C*-terminal carboxylate. In the flanking regions of the cluster are other genes that encode enzymes that are involved in the biosynthesis of some of the unusual amino acids and fatty acids building blocks, along with enzymes that are responsible for tailoring of the nascent peptide [11] (Fig. 3).

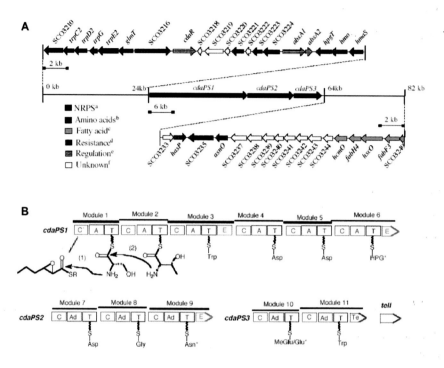

Figure 3. (A) Organization of the CDA biosynthetic gene cluster. **(B)** The CDA nonribosomal peptide synthase (NRPS).

WILD-TYPE CDAS WITH NOVEL Z-DEHYDROTRYPTOPHAN RESIDUES

Earlier work [12] had revealed the structures of four wild-type calcium-dependent antibiotics, CDA1b, 2b, 3b and 4b (Fig. 2), which differ in the substitution pattern at amino acid residues 9 and 10. Small amounts of other peptides were also isolated [12] but these were not characterized. Initially we set out to optimize the wild-type CDA production levels and to isolate and characterize any new natural CDA variants. Accordingly *S. coelicolor* strains 2377 [21] and MT1110 [21] were cultivated in a variety of different media and the supernatants where subsequently analysed by LC-MS. This revealed a similar profile of CDAs to that described previously [12]. On the other hand when strain 2377 was grown on solid media, Oxoid Nutrient Agar (ONA) LC-MS of the exudates revealed two previously uncharacterized CDA2a (1574 Da) and CDA4a (1494 Da) as the major products (Fig. 2).

Following large-scale solid media cultures and purification (HPLC) sufficient quantities of these peptides where isolated, which allowed a detailed structural elucidation using UV, NMR and mass spectrometry (MS). The structures of these metabolites differ from the b-series of CDA by the presence of an unusual Z-2,3-dehydrotryptophan (ΔTrp) residue rather than L-tryptophan. As there are two Trp derived residues in CDA at position 3 and 11, it was necessary to establish which of the two had been oxidized. This was achieved by ring opening CDA4a to generate a linear peptide, which was sequenced by tandem mass spectrometry. This clearly showed that the ΔTrp residue was at the C-terminus. Notably there are only a few other known nonribosomal peptides, including keramamide F and the microsclerodermins that contain ΔTrp [23, 24]. The structure of CDA3a (1480 Da) was also assigned (Fig. 2) on the basis of these findings.

MUTASYNTHESIS OF CDA WITH MODIFIED ARYLGLYCINE RESIDUES

In the cda gene cluster there are three genes hmaS (SCO3229), hmo (SCO3228), and hpgT (SCO3227) that encode proteins that display similarity to enzymes encoded by genes from the vancomycin-type antibiotic biosynthetic gene clusters in Amycolatopsis orientalis [25] and Streptomyces lavendulae [26]. These gene products are known to be involved in the biosynthesis of L-4-hydroxyphenylglycine (L-HPG) 3 [27,28]. HmaS catalyses the oxidative decarboxylation of 4-hydroxyphenylpyruvate to give L-4-hydroxymandelic acid 1 (Fig. 4). This then is oxidized by Hmo to give 4-hydroxyphenylglyoxylate 2, which is finally transaminated by HpgT resulting in L-HPG 3.

An initial attempt to engineer the biosynthesis of new CDA lipopeptides focused on the L-HPG pathway and utilized a mutasynthesis approach to generate new CDAs with modified arylglycine residues at position 6. Accordingly the first gene in the L-HPG pathway hmaS was deleted from the CDA cluster by standard 'double crossover' gene replacement. The resulting S. coelicolor mutant (ΔhmaS), is deficient in L-HPG 3 and the precursors 1 and 2 (Fig. 4) and was unable to produce CDA, as determined by both bioassays and LC-MS analysis. Feeding the mutant 4-hydroxymandelate 1,4-hydroxyphenylglyoxylate 2 or L-HPG 3 re-establishes the production of CDA in liquid and on solid culture media. This clearly demonstrates that, as expected, the pathway responsible for the biosynthesis L-HPG in A. orientalis [27,28] also operates in S. coelicolor.

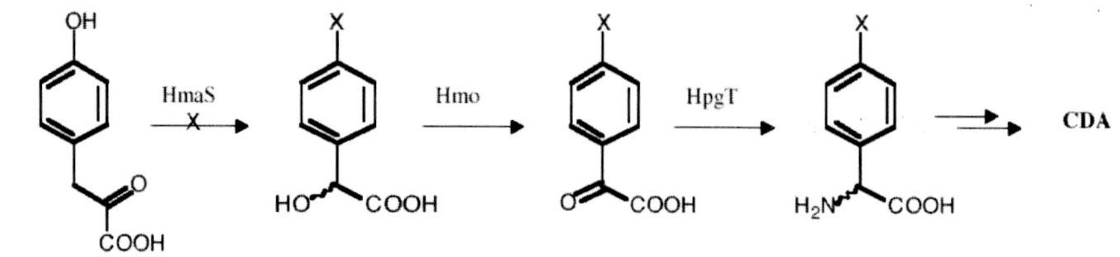

4-Hydroxy-phenylpyruvate

X			
X = OH	**1**→ CDA2b	**2**→ CDA2b	**3** (L)→ CDA2b
			3 (D)→ no CDA
H	**4**→ CDA2d	**5**→ CDA2d	**6**→ CDA2d
F	**6**→ no CDA	**8**→ CDA2fa & 2fb	**9**→ CDA2fa & 2fb
Cl	**10**→ no CDA	**11**→ no CDA	**12**→ no CDA
OCH_3	**13**→ no CDA	**14**→ no CDA	**15**→ no CDA

Figure 4. Mutasynthesis of CDA with modified arylglycine residues. HmaS = 4-hydroxymandelate synthase, Hmo = 4-hydroxymandelate oxidase, HpgT = 4-hydroxyphenylglycine transaminase. Both D- and L- enantiomers of HPG **3** were fed only the L-enantiomer is incorporated. All the other substrates were racemic.

Moreover, feeding 4-dehydroxy (4, 5, & 6) and 4-fluoro (8 &9) analogues of these precursors to the Δ*hmaS* mutant, grown in liquid culture media, resulted in new CDA peptides CDA2d, possessing modified phenylglycine (R_6 = H), as well as CDA2fa and CDA2fb with a 4-fluorophenylglycine (R_6 = F) (Fig. 4). The structures of these new peptides were confirmed by extensive NMR and mass spectrometry experiments [11]. In contrast 4-chloro 10-12 and 4-methoxy 13-16 analogues did not give rise to detectable CDA by LC-MS analysis. Presumably increasing the size of the C4-substituent, beyond the size of an hydroxyl group, results in failure of the NRPS L-HPG-activating A-domain to recognize and activate the modified arylglycines. This work is significant as it represents the first rational engineered biosynthesis of acidic lipopeptides antibiotics of this class. In addition this strategy could also be used to modify other important HPG-containing non-ribosomal peptides, such as those from the vancomycin group of antibiotics [29](Fig. 1).

Site-Directed Mutagenesis of Adenylation Domains

The next approach to engineer new CDAs focused on active site modification of the NRPS A-domains. Previously the 10 key residues at the active site of the A-domain that are responsible for binding the amino acid substrate were identified, using the X-ray crystal structure of (PheA), a Phe-activating A-domain, from gramicidin S synthetase [30,31].

Subsequently *in vitro* studies with recombinant A-domains showed that by changing as few as one of these residues it was possible to alter the specificity of the A-domain from activating one amino acid to another [30]. It was therefore anticipated, although at the outset unproven, that this approach might be extend *in vivo* in order to generate nonribsomal peptides with different amino acid sequences.

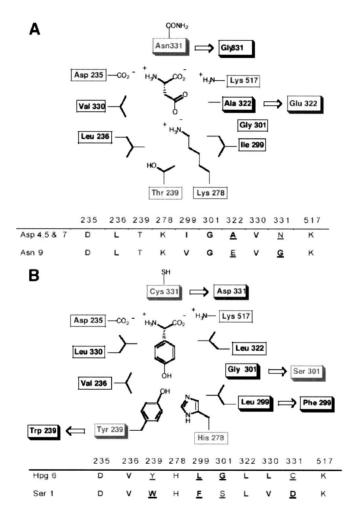

Figure 5. The active site architecture of **(A)** the module 7 Asp-activating A-domain and **(B)** the module 6 L-HPG activating A-domain. The residues of the amino acid binding pockets were determined through alignment with PheA [30, 31]. The tables below show the alignments of the Asp- and HPG-activating A-domains with the Asn and Ser activating A-domains of modules 9 and 1 respectively.

Alignment of the CDA NRPS A-domains with PheA enabled the key active site residues of each A-domain to be identified. Noticeably the active sites of the Asp-activating A-domains of modules 4, 5 and 7 are all identical to each other and also similar to the module 9 Asn activating A-domain differing at positions 299, 322 and 331 (Fig. 5A). Given that the residues at position 299 are similar (Val vs Ile) it was argued that changing Ala322→Glu and Asn331→Gly at one of the Asp-activating A-domains should result in a the incorporation of Asn into CDA instead of Asp. Accordingly, a DNA fragment encompassing module 7, was subjected to site directed mutagenesis to generate single (Ala322→Glu) and double-point (Ala322Æ Glu, Asn331→Gly) mutants which where cloned into a delivery plasmid to give pGUM7S and pGUM7D respectively. These plasmids were used to deliver the mutations to the appropriate regions of *cdaPS2* on the *S. coelicolor* chromosome through homologous recombination. The single-point mutant GUM7S failed to produce any CDA with Asn or Asp at position 7. A new product was however identified by LC-MS which was isolated and subsequently shown to the CDA-hexapeptide intermediate **16** by high resolution MS (m/z 884.3321 [M+H]$^+$, $C_{40}H_{50}N_7O_{16}$ requires 884.3314) and tandem MS (Fig. 6A). On the other hand the double-point mutant GUM7D produced both the linear 6mer **16** and a new product which exhibits m/z 1574.30 [M+H]$^+$, which is consistent with a new lipopeptide, CDA2a-7N (Fig. 2) possessing Asn at position 7 rather than Asp, that is not seen in the wild-type. Significantly, extracts containing CDAa-7N are inactive in bioassays, which suggests that Asp-7 is essential for calcium binding and antibiotic activity of CDA.

During this study a similar approach was used to generate modified variants of the lipopeptide surfactin in *Bacillus subtilis* [10]. In this case no peptidyl intermediates where identified. Therefore to test the generality of our findings another *S. coelicolor* mutant GUM6S was generated possessing a single-point mutation (Gly301→Ser) of the module 6 HPG-activating A-domain (Fig. 5B). It is predicted [31] that this mutation should change the specificity of the A-domain from HPG to Ser. LC-MS analysis of the mutant failed to show the production of CDA with Ser at position 6. However a new product was identified which corresponds in m.w. (734 Da) and HPLC retention time to the CDA 5mer intermediate 2,3-epoxyhexanoyl-Ser-Thr-D-Trp-Asp-Asp-OH 17. A module 6, A-domain quadruple point mutant (Gly301→Ser, Leu299→Phe, Cys331→Asp & Tyr239→Trp) was also prepared but this also failed to generate CDA with Ser at position 6 instead of HPG.

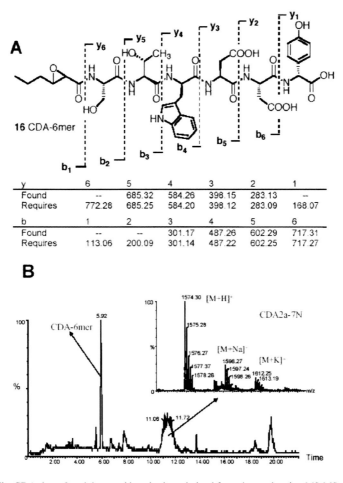

y	6	5	4	3	2	1
Found	--	685.32	584.26	398.15	283.13	--
Requires	772.28	685.25	584.20	398.12	283.09	168.07

b	1	2	3	4	5	6
Found	--	--	301.17	487.26	602.29	717.31
Requires	113.06	200.09	301.14	487.22	602.25	717.27

Figure 6. (A) The CDA-6mer **3** and the y and b series ions derived from the product ion MS-MS spectra. **(B)** LC-MS of extracts from the *S. coelicolor* double-point mutant GUM7D.

EVIDENCE OF THE EXISTENCE OF AN HITHERTO ELUSIVE
NRPS PROOF-READING MECHANISM

It is possible, given the lability of the peptidyl thioester bond, that non-enzymatic hydrolysis could have caused the release of the CDA-6mer **16** and 5mer **17**. However, the clear absence of any of the other shorter intermediates in the culture supernatant means this is unlikely. It is more likely that changes to the active site of the A-domain result in failure or less efficient activation of non-cognate amino acids, which brings about a kinetic blockage on the NRPS that signals the enzymatic hydrolysis of the stalled upstream peptidyl chain.

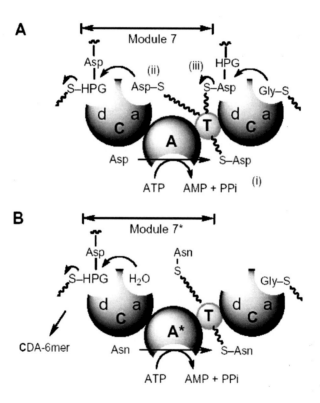

Figure 7. (A) A schematic description of the reactions catalysed by the module 7 of cdaPS2. (i) The A-domain activates and transfers the substrate amino acid to the Ppant side chain of the T-domain. (ii) The thioester intermediate translocates to the acceptor (a) site on the upstream C-domain, where it intercepts the upstream hexapeptidyl thioester intermediate in the donor (d) site. (iii) The resulting heptapeptidyl thioester is then translocated to the d-site on the down stream C-domain, before formation of the next peptide bond with Gly. **(B)** The mutant module 7 A-domain (GUM7D) activates Asn instead of Asp. However, the Asn-thioester intermediate is only weakly recognized by the upstream C-domain a-site. A water molecule can thus compete for the a-site and intercept the hexapeptidyl intermediate in the d-site.

In the *cda* gene cluster, down stream of *cdaPS3*, is gene *tell* that encodes a typical type II thioesterase (Fig. 3). It has been postulated that these type II thioesterases may be involved in proof-reading NRPS [32] and relate modular polyketide synthases (PKS) [33, 34]. A double mutant GUM7S-Δ*tell* was therefore generated where the *tell* gene was deleted from the *S. coelicolor* module 7, A-domain single-point mutant (GUM7S). This was also shown to produce the CDA-6mer 16, which suggests that TelI is not responsible for the hydrolysis of the stalled peptidyl intermediates. Instead we have proposed [35] that an alternative NRPS proof-reading mechanism exists where the upstream condensation (C) domain catalyses the hydrolysis of the stalled peptidyl intermediates (Fig. 7). According to this hypothesis a modified A-domain activates a non-cognate amino acid substrate, which is transferred to the phosphopantetheine group of the thiolation (T) domain. The amino acyl-*S*-Ppant group then translocates to the acceptor (a) site [36] on the upstream C-domain. However, the a-site is unable to, or only weakly able to, recognize the non-cognate substrate. Consequently a water molecule can bind to the a-site, then using the catalytic functionality of the C-domain, intercept the peptidyl-thioester intermediate in the donor (d) site of the C-domain. As a result of these findings we suggest that to incorporate a non-cognate amino acid into a nonribosomal peptide efficiently, it is likely that the specificity of the upstream C-domain acceptor site must be altered, as well as the A-domain.

CONCLUSION

We have isolated and elucidated the structures of several previously uncharacterized CDAs, which possess unusual *Z*-dehydrotryptophan residues at the C-terminus. In addition we were first to engineer the biosynthesis of novel lipopeptides of the CDA/daptomycin class. Our first approach involved a mutasynthesis strategy. Here a gene involved in the biosynthesis of L-hydroxyphenylglycine (L-HPG) was deleted. Analogues of HPG and its precursors were then fed to the resultant mutant to give several novel engineered CDAs with modified arylglycine residues at position 6. We also showed that site directed mutagenesis of the module 7 A-domain leads to a new engineered lipopeptide product (CDA2a-7N) with Asn at position 7 instead of Asp which is found in the wild type CDA. However the levels of CDA2a-7N were considerably reduced compared with the wild-type CDA. This was due to the premature hydrolysis of upstream peptidyl intermediate.

These findings are significant because they point to the existence of a hitherto elusive NRPS proof-reading mechanism, which must be further elucidated and eventually circumnavigated, if this approach is to be successful in delivering the quantities of modified products required for drug development.

ACKNOWLEDGEMENTS

The work was supported by the BBSRC (grant 36/B12126), BBSRC/GlaxoSmithKline and BBSRC/Biotica Technologies through the award of PhD CASE studentships. We especially thank all our co-workers who are named in the references to their original work that are shown below.

REFERENCES

[1]   Schwazer, D., Finking, R., Marahiel, M.A. (2003) Nonribosomal peptides: From genes to products. *Nat. Prod. Rep.* **20**:275-287.

[2]   Marahiel, M.A., Stachelhaus, T., Mootz, H.D. (1997) Modular peptide synthetases involved in nonribosomal peptide synthesis. *Chem. Rev.* **97**:2651-2673.

[3]   Mootz, H.D., Schwarzer, D., Marahiel, M.A. (2002) Ways of assembling complex natural products on modular nonribosomal peptide synthetases. *ChemBioChem.* **3**:490-504.

[4]   Walsh, C.T. (2002) Combinatorial biosynthesis of antibiotics: challenges and opportunities. *ChemBioChem.* **3**:124-134.

[5]   Cane, D.E., Walsh, C.T., Khosla, C. (1998) Harnessing the biosynthetic code: Combinations, permutations, and mutations. *Science* **282**:63-68.

[6]   Doekel, S., Marahiel, M.A. (2001) Biosynthesis of natural products on modular peptide synthetases. *Metabol. Engng* **3**:64-77.

[7]   Walsh, C.T. (2004) Polyketide and nonribosomal peptide antibiotics: Modularity and versatility. *Science* **303**:1805.

[8]   Stachelhaus, T., Schneider, A., Maraheil, M.A. (1995) Rational design of peptide antibiotics by targeted replacement of bacterial and fungal domains. *Science* **269**:69-72.

[9]   Mootz, H.D., Kessler, N., Linne, U., Eppelmann, K., Schwarzer, D., Marahiel, M.A. (2002) Decreasing the ring size of a cyclic nonribosomal peptide antibiotic by in-frame module deletion in the biosynthetic genes. *J. Am. Chem. Soc.* **124**:10980-10981.

[10]   Eppelmann, K., Stachelhaus, T., Marahiel, M.A. (2002) Exploitation of the selectivity-conferring code of nonribosomal peptide synthetases for the rational design of novel peptide antibiotics. *Biochemistry* **41**:9718-9726.

[11]   Hojati, Z., Milne, C., Harvey, B., Gordon, L., Borg, M., Flett, F., Wilkinson, B., Sidebottom, P.J., Rudd, B.A.M., Hayes, M.A., Smith C.P., Micklefield, J. (2002) Structure, biosynthetic origin, and engineered biosynthesis of calcium-dependent antibiotics from *Streptomyces coelicolor. Chem. Biol.* **9**:1175-1187.

[12]   Kempter, C., Kaiser, D., Haag, S., Nicholson, G., Gnau, V., Walk, T., Gierling, G.H., Decker, H., Zähner, H., Jung, G., Metzger, J.W. (1997) CDA: Calcium-dependent peptide antibiotics from *Streptomyces coelicolor* A3(2) containing unusual residues. *Angew. Chem. Int. Ed. Engl.* **36**:498-501.

[13]   Ball, L.-J., Goult, C.M., Donarski, J.A., Micklefield J., Ramesh, V. (2004) NMR structure determination and calcium binding effects of lipopeptide antibiotic daptomycin. *Org. Biomol. Chem.* **2**:1872-1878.

[14]   Debono, M., Abbott, B.J., Molloy, R.M., Fukuda, D.S., Hunt, A.H., Daupert, V.M., Counter, F.T., Ott. J.L., Carrell, C.B., Howard, L.C., Boeck, L.D., Hamill, R.L. (1988) Enzymatic and chemical modifications of lipopeptide antibiotic A21978C: The synthesis and evaluation of daptomycin (LY146032). *J. Antibiotics* **41**:1093-1105.

[15]   Fukuda, D.S., Du Bus, R.H., Baker, P.J., Berry, D.M., Mynderse, J. S. (1990) A54145, a new lipopeptide antibiotic complex: isolation and characterization. *J. Antibiotics* **43**:594-615.

[16]   Vértesy, L., Ehlers, E., Kogler, H., Kurz, M., Meiwes, J., Seibert, G., Vogel, M., Hammann, P. (2000) Friulimicins: Novel lipopetide antibiotic with peptidoglycan synthesis inhibiting activity from *Actinoplanes friuliensis* sp. nov. *J. Antibiotics* **53**:816-827.

[17]   Raja, A., LaBonte, J., Lebbos J., Kirkpatrick, P. (2003) Daptomycin. *Nature Rev. Drug Discov.* **2**:943-944.

[18]   Micklefield, J. (2004) Daptomycin structure and mechanism of action revealed. *Chem. Biol.* **11**:887-895.

[19]   Chong, P.P., Podmore, S.M., Kieser, H.M., Redenbach, M., Turgay, K., Marahiel, M.A., Hopwood, D.A., Smith, C.P. (1998) Physical identification of a chromosomal locus encoding biosynthetic genes for the lipopeptide calcium-dependent antibiotic (CDA) of *Streptomyces coelicolor* A3(2). *Microbiology* **144**:193-199.

[20]   Bentley, S.D. *et al.* (2002) Complete genome sequence of the model actinomycete *Streptomyces coelicolor* A3(2). *Nature* **417**:141-147.

[21]   Hopwood, D.A., Wright, H.M. (1983) CDA is a new chromosomally-determined antibiotic from *Streptomyces coelicolor* A3(2). *J. Gen. Microbiol.* **129**:3575-3579.

[22]   Hindle, Z., Smith, C.P. (1994) Substrate induction and catabolite repression of the *Streptomyces coelicolor* glycerol operon are mediated through the GylR protein. *Mol. Microbiol.* **12**:737-745.

[23] Itagaki F., Shigemori, H., Ishibashi, M., Nakamura, T., Sasaki, T., Kobayashi, J. (1992) Keramamide F, a new thiazole-containing peptide from the Okinawan marine sponge *Theonella* sp. *J. Org. Chem.* **57**:5540-5542.

[24] Quershi, A., Colin, P.L., Faulkner, D.J. (2000) Microsclerodermins F-I, antitumor and antifungal cyclic peptides from the lithistid sponge *Microscleroderma* sp. *Tetrahedron* **56**: 3679-3685.

[25] van Wageningen, A.M.A., Kirkpatrick, P.N., Williams, D.H., Harris, B.R., Kershaw, J.K., Lennard, N.J., Jones, M., Jones, S.J.M., Solenberg, P.J. (1998) Sequencing and analysis of genes involved in the biosynthesis of a vancomycin group antibiotic. *Chem. Biol.* **5**:155-162.

[26] Chiu, H.-T., Hubbard, B.K., Shah, A.N., Eide, J., Fredenburg, R.A., Walsh, C.T., Khosla, C. (2001) Molecular cloning and sequence analysis of the complestatin biosynthetic gene cluster. *Proc. Natl. Acad. Sci. USA* **98**:8548-8553.

[27] Choroba, O.W, Williams, D.H., Spencer, J.B. (2000) Biosynthesis of the vancomycin group of antibiotics: involvement of an unusual dioxygenase in the pathway to (*S*)-4-hydroxyphenylglycine. *J. Am. Chem. Soc.* **122**:5389-5390.

[28] Hubbard, B.K., Thomas, M.G., Walsh, C.T. (2000) Biosynthesis of L-*p*-hydroxyphenylglycine, a non-proteinogenic amino acid constituent of peptide antibiotics. *Chem. Biol.* **7**:931-942.

[29] Weist, S., Bister, B., Puk, O., Bischoff, D., Pelzer, S., Nicholson, G.J., Wohlleben, W., Jung, G., Süßmuth, R.D. (2002) Fluorobalhimycin - a new chapter in glycopeptide research. *Angew. Chem. Int. Ed. Engl.* **41**:3383-3385.

[30] Stachelhaus, T., Mootz, H.D., Marahiel, M.A. (1999) The specificity-conferring code of adenylation domains in nonribosomal peptide synthetases. *Chem. Biol.* **6**:493-505.

[31] Challis, G.L., Ravel, J., Townsend, C.A. (2000) Predictive, structure-based model of amino acid recognition by nonribosomal peptide synthetase adenylation domains. *Chem. Biol.* **7**:211-224.

[32] Schwarzer, D., Mootz, H.D., Linne, U., Marahiel, M.A. (2002) Regeneration of misprimed nonribosomal peptide synthetases by type II thioesterases. *Proc. Natl. Acad. Sci. USA* **99**:14083-14088.

[33] Heathcote, M.L., Staunton, J., Leadley, P.F. (2001) Role of type II thioesterases: evidence for removal of short acyl chains produced by aberrant decarboxylation of chain extender units. *Chem. Biol.* **8**:207-220.

[34] Kim, B.S., Cropp, T.A., Beck, B.J., Sherman, D.H., Reynolds, K.A. (2002) Biochemical evidence for an editing role of thioesterase II in the biosynthesis of the polyketide pikromycin. *J. Biol. Chem.* **277**:48028-48034.

[35] Uguru, G. C., Milne, C., Borg, M., Flett, F., Smith, C.P., Micklefield, J. (2004) Active-site modifications of adenylation domains lead to hydrolysis of upstream nonribosomal peptidyl thioester intermediates. *J. Am. Chem. Soc.* **126**:5032-5033.

[36] Belshaw, P.J., Walsh, C.T., Stachelhaus, T. (1999) Aminoacyl-CoAs as probes of condensation domain selectivity in nonribosomal peptide synthesis. *Science* **284**:486-489.

CHEMICAL COMPLEMENTATION:
A REACTION-INDEPENDENT GENETIC ASSAY FOR
DIRECTED EVOLUTION

VIRGINIA W. CORNISH*, HENING LIN, KATHLEEN BAKER,
GILDA J. SALAZAR-JIMENEZ, WASSIM ABIDA, COLLEEN BLECKZINSKY,
DEBLEENA SENGUPTA, SONJA KRANE AND HAIYAN TAO

Department of Chemistry, Columbia University, New York, NY 10027, U.S.A.
E-Mail: *vc114@columbia.edu

Received: 12th July 2004 / Published: 22nd July 2005

INTRODUCTION

Enzymes are able to catalyse a broad range of chemical transformations not only with impressive rate enhancements but also with both regio- and stereo-selectivity and so are attractive candidates as practical alternatives to traditional small molecule catalysts. With applications as diverse as chemical synthesis, reagents for commercial products and biomedical research, and even therapeutics, there is a great demand for enzymes with both improved activity and novel catalytic function [1,2]. In theory, the properties of an enzyme can be altered by rational design; however, rational design is greatly hindered in practice by the complexity of protein function. With advances in molecular biology the possibility has arisen that an enzyme with the desired catalytic property can instead be isolated from a large pool of protein variants. Recently directed evolution has been used successfully to modify the substrate [3] or cofactor specificity [4] of an existing enzyme. These experiments, however, are limited to reactions that are inherently screenable or selectable-reactions where the substrate is a peptide [6,7] or the product is fluorescent [4] or an essential metabolite [3,8,9]. What is needed now to realize the power of directed evolution experiments are screening and selection strategies that are general-strategies that do not limit the chemistry and that can readily be adapted to a new target reaction.

Early success with assays based on binding to transition-state analogues and suicide substrates convinced researchers that it should be possible to engineer proteins to catalyse a broad range of reactions [10], but it was difficult to translate binding events into read-outs for enhanced catalytic activity. Recently, attention has turned to direct selections for catalytic activity. While strategies ranging from *in vitro* fluorescence assays to physically linking the enzyme to its substrate have all recently been reported [10-15], a general solution to this problem is yet to emerge. *In vivo* complementation, in which an enzyme is selected based on its ability to complement an essential activity that has been deleted from a wild-type cell, has proven to be one of the most powerful approaches to enzyme evolution [3,5,8]. However, complementation assays are limited to natural reactions that are selectable. In this paper we describe our efforts to develop a "chemical complementation" system, which would allow us to control the chemical reaction linked to cell survival, extending complementation approaches to a broad range of chemical transformations.

RESULTS AND DISCUSSION

Strategy

Our approach to developing such a chemical complementation system is to use the yeast three-hybrid assay to link enzyme catalysis to transcription of a reporter gene *in vivo* (Fig. 1).

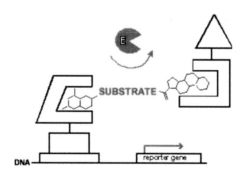

Figure 1. Chemical Complementation. A reaction-independent complementation assay for enzyme catalysis based on the yeast three-hybrid assay. A heterodimeric small molecule bridges a DNA-binding domain-receptor fusion protein and an activation domain-receptor fusion protein, activating transcription of a downstream reporter gene *in vivo*. Enzyme catalysis of either cleavage or formation of the bond between the two small molecules can be detected as a change in transcription of the reporter gene. The assay can be applied to new chemical reactions simply by synthesizing small molecules with different substrates as linkers and adding an enzyme as a fourth component to the system.

As a first step, we designed a yeast three-hybrid system in which a dimeric small molecule bridges a DNA-binding domain-receptor fusion protein and an activation domain-receptor fusion protein. By dimerizing the two fusion proteins via the receptors, the small molecule effectively reconstitutes the transcriptional activator, turning on transcription of a downstream reporter gene. We envisioned that this system could be used as a read-out for enzyme catalysed bond cleavage or formation reactions simply by replacing the chemical linker between the two small molecules with the bond to be cleaved or formed and adding an enzyme as a fourth component to the system. Cleavage of the bond between the two small molecules by an active enzyme would disrupt reconstitution of the transcriptional activator and, hence, transcription of the reporter gene. Bond formation would activate transcription of the reporter gene. This approach is general because it can be applied to new reactions simply by synthesizing heterodimers with different substrates as chemical linkers. Genetic assays are high-throughput because large pools of proteins can be sorted based on clear changes in cellular phenotype. We then developed the chemical complementation system around the well-studied reaction, cephalosporin hydrolysis by a cephalosporinase (β-lactam hydrolase, EC 3.5.2.6). Using this model system, we went on to show that the catalytic efficiency (k_{cat}/K_M) of cephalosporinase variants correlates with their levels of transcription activation in the three-hybrid assay. In preliminary results, we have now demonstrated the generality of this approach, applying it to the directed evolution of an enzyme that catalyses glycosidic bond formation.

Three-Hybrid System [16]

The first step then was to design a three-hybrid system in which dimerization of a transcriptional activator is dependent on a heterodimeric small molecule. For this work, we built from a dexamethasone-FK506 yeast three-hybrid system reported by Licitra and Liu [17] and precedent from the Schreiber and other laboratories in the design of chemical inducers of dimerization (CIDs) [18,19]. We chose to build a heterodimeric CID based on the well-characterized ligand-receptor pairs dexamethasone (Dex)-glucocorticoid receptor (GR) and methotrexate (Mtx)-dihydrofolate reductase (DHFR). Both Dex and Mtx present chemical functionality that can be modified readily without disrupting receptor binding. The rat glucocorticoid receptor (rGR) binds Dex with a KD of 5 nM, and mutants of rGR with increased affinity for Dex have been isolated [20].

The steroid dexamethasone has been used extensively as a cell-permeable small molecule to regulate the activity and nuclear localization of GR fusion proteins *in vivo* [21]. Dex is commercially available and can be derivatized at the C-21 position without disrupting its binding to GR [22,23]. Likewise, the interaction between Mtx and DHFR is extremely well characterized both biochemically and structurally [24,25]. DHFR fusion proteins have been used for a variety of biochemical applications due to Mtx's picomolar K_I for DHFR [26]. Mtx is cell permeable, commercially available, and can be synthesized readily from simple starting materials.

Scheme 1

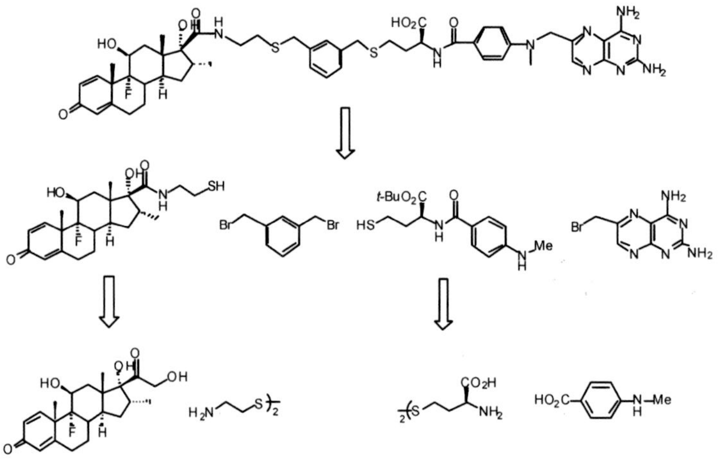

The retro-synthetic analysis of the Dex-Mtx heterodimer is shown in Scheme 1. The synthesis is based on previous syntheses of Dex and Mtx derivatives [22,23]. The synthesis is designed to allow the chemical linker between the two ligands to be varied readily. Both ligands were introduced as thiol derivatives to a di-halo linker. Following oxidative cleavage with periodate, Dex was dervatized with cystamine using standard peptide coupling reagents.

The γ-carboxylate in Mtx was replaced with a thiol simply by replacing glutamate with homocysteine. Homocysteine, protected as the tert-butyl ester and disulfide, was coupled to 4-methylaminobenzoic acid. The resulting Dex and Mtx disulfide derivatives were reduced to their corresponding thiols using tri-n-butylphosphine, and the two thiols were coupled to a di-bromo linker in a one-pot reaction. The 2,4-diamino-6-bromomethyl pteridine was added after introduction of the dibromo linker to simplify purification of the intermediates, and the final step was cleavage of the tert-butyl ester.

Using a modification of this original synthesis, the Dex-Mtx heterodimer is now routinely prepared from two components in 6 steps in 10% overall yield. The modular synthesis of the Dex-Mtx heterodimer was designed with the chemical complementation system in mind, where the linker between the two halves of the molecule will need to be varied.

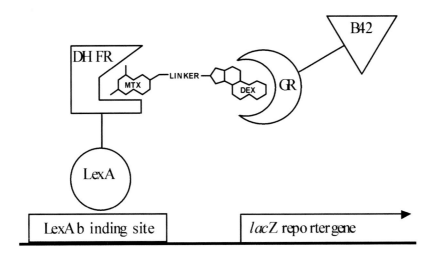

Figure 2. Yeast three-hybrid assay. A binding site with high affinity for the DNA-binding protein LexA is placed upstream of a *lacZ* reporter gene, which encodes β-galactosidase. The LexA-DHFR protein chimera binds to the LexA binding site. In the presence of the Dex-Mtx CID, the LexA-DHFR chimera binds to Dex-Mtx which in turn binds to the GR-B42 protein chimera. The Dex-Mtx CID, thus, effectively brings the transcription activation protein B42 to the LexA binding site. Once bound to the LexA binding site, B42 recruits the transcriptional machinery to the *lacZ* reporter gene, and β-galactosidase is synthesized. The levels of β-galactosidase synthesis, therefore, are a good indicator of the efficiency of Dex-Mtx-induced protein dimerization.

For the genetic assay, we chose the yeast "three-hybrid" system [17,28,29], which consists of a dimeric small molecule, two protein chimeras, and a reporter gene (Fig. 2). The small molecule (Dex-Mtx) bridges the DNA-binding protein chimera (LexA-DHFR) and the transcription activation protein chimera (B42-GR). Dimerization of LexA and B42 recruits the transcription machinery to the promoter near the LexA-binding site, thereby activating transcription of the *lacZ* reporter gene. Plasmids encoding the LexA-DHFR and B42-GR protein chimeras were prepared from the Brent two-hybrid vectors pMW102 and pMW103 using standard molecular biology techniques [30,31]. Plasmid pMW106, which encodes the *lacZ* reporter gene under control of four tandem LexA operators, was used as provided. All three plasmids were introduced into *S. cerevisiae* strain FY250, and the resultant strain was used to test the activity of the Dex-Mtx CID.

Cornish, V. *et al.*

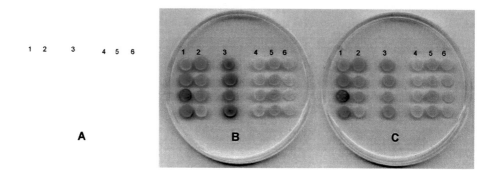

Figure 3. X-gal plate assay of Dex-Mtx-induced *lacZ* transcription. Yeast strains containing a *lacZ* reporter gene and different LexA- and/or B42-chimeras were grown on X-gal indicator plates with or without Dex-Mtx. Columns **1-6** on each plate correspond to yeast strains containing different LexA- and/or B42-chimeras: **1)** LexA-Sec16p, B42-Sec6p; **2)** LexA-Sec13, B42-Sec6p. 1 and 2 are direct protein-protein interactions used as positive controls [27]; **3)** LexA-DHFR, B42-GR; **4)** LexA-DHFR, B42; **5)** LexA, B42-GR; **6)** LexA, B42. X-gal plates **A-C** have different small molecule combinations: **A)** no Dex-Mtx; **B)** 1 μM Dex-Mtx; **C)** 1 μM Dex-Mtx, and 10 μM Mtx.

Using standard b-galactosidase activity assays both on plates and in liquid culture [31] we showed that Dex-Mtx could activate *lacZ* transcription *in vivo* (Fig. 3 and data not shown). Based on previous studies showing that *lacZ* transcription levels correlate with the strength of protein-protein interactions in the yeast two-hybrid assay, we expect β-galactosidase activity to be a good indicator of Dex-Mtx induced protein dimerization. In these assays, the extracellular concentration of Dex-Mtx ranged from 0.01 to 10 μM. No activation was observed at concentrations < 0.1 μM. Control experiments established that *lacZ* transcription was dependent on Dex-Mtx (Fig. 3 and data not shown). Only background levels of β-galactosidase activity were detected when Dex-Mtx was omitted. A 10-fold excess of Mtx reduced Dex-Mtx-dependent *lacZ* transcription to near background levels. A 10-fold excess of Dex, however, did not affect Dex-Mtx-dependent *lacZ* transcription, and higher concentrations of Dex were toxic to the yeast cells.

This result may be due to differences in cell permeability between Dex and Dex-Mtx or may suggest that LexA-DHFR, but not B42-GR, is the limiting reagent. To provide further support that *lacZ* transcription is Dex-Mtx-dependent, LexA-DHFR was replaced with LexA, or B42-GR was replaced with B42. Again, only background levels of *lacZ* transcription were detected in the presence of Dex-Mtx in the yeast cells containing either LexA and B42-GR or LexA-DHFR and B42.

The Dex-Mtx yeast three-hybrid system proved quite robust, providing the framework for subsequent development of the chemical complementation system. The Dex-Mtx heterodimer could be synthesized readily and cheaply, facilitating future incorporation of the substrate bond into this molecule. The Dex-Mtx heterodimer is a strong transcriptional activator, which is essential in the next stage for linking enzyme catalysis to reporter gene transcription. Because the three-hybrid system is the cornerstone for the chemical complementation system, our laboratory has invested significantly in the development and optimization of this assay. We have characterized the system and made improvements [39], notably the development of a bacterial three-hybrid system, which ultimately should allow large libraries to be evaluated because of the high transformation efficiency of *E. coli* [40].

Chemical Complementation [32]

With the Dex-Mtx yeast three-hybrid system in hand, we next set out to develop the chemical complementation system. We chose cephalosporin hydrolysis by the *Enterobacter cloacae* P99 cephalosporinase as a simple cleavage reaction to demonstrate the selection strategy (Fig. 4). Cephalosporins are β-lactam type antibiotics, and cephalosporinases are the bacterial resistance enzymes that hydrolyse and, therefore, inactivate these antibiotics. The cephalosporinase enzyme is well-characterized biochemically and structurally [33,34], and the synthesis of cephem compounds established [35]. We chose to incorporate Mtx and Dex at the C3' and C7 positions, respectively, of the cephem core. Cleavage of the β-lactam bond in cephalosporins results in expulsion of the leaving group at the C3' position, effectively breaking the bond between Mtx and Dex (Fig. 4A). Thus, the Mtx-Cephem-Dex substrate should dimerize the transcriptional activator causing transcription of the reporter gene in the yeast three-hybrid assay. When the cephalosproinase enzyme is expressed, however, the cephem linkage should be cleaved, and protein dimerization and transcription of the reporter gene should be disrupted.

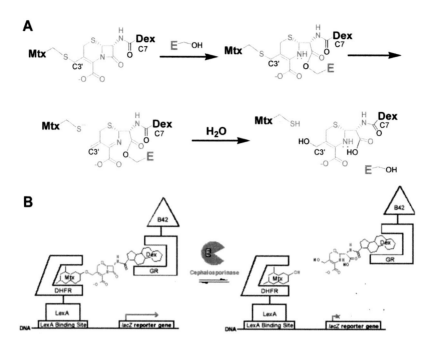

Figure 4. Cephalosporinase Model Reaction. Cephalosporin hydrolysis provides a simple cleavage reaction to demonstrate the complementation strategy. **A)** Cephalosporin hydrolysis by a cephalosporinase enzyme. Cephalosporinases are serine-protease type enzymes and catalyze the hydrolysis of cephalosporin antibiotics. Hydrolysis of the β-lactam bond in Mtx-Cephem-Dex results in expulsion of the leaving group at the C3' position of the cephem core, effectively breaking the bond between Mtx and Dex. **B)** Cephalosporin hydrolysis by the cephalosporinase enzyme disrupts transcription of a *lacZ* reporter gene. The Mtx-Cephem-Dex substrate dimerizes a LexA DNA-binding domain-dihydrofolate reductase (LexA-DHFR) and a B42 activation domain-glucocorticoid receptor (B42-GR) fusion protein, activating transcription of a *lacZ* reporter gene. Addition of active cephalosporinase enzyme results in cleavage of the Mtx-Cephem-Dex substrate and disruption of *lacZ* transcription.

While previous work with the yeast two- and three-hybrid assays suggests that the linker between Mtx and Dex should not affect the transcription read-out, we first wanted to test our assumption that Mtx-Cephem-Dex would retain the ability to activate transcription in our yeast three-hybrid system. In addition, we wanted to carry out *in vitro* kinetic studies to ensure that Mtx-Cephem-Dex is an efficient substrate for the cephalosporinase enzyme. The first step was to synthesize the Mtx-Cephem-Dex substrate. The original synthesis of the Mtx-Dex heterodimer had been designed to facilitate the inclusion of different chemical linkers between the Mtx and Dex portions of the molecule [16]. Thus, the commercial cephem intermediate, 7-amino-3-chlormethyl-3-cephem-4-carboxylic acid p-methoxybenzyl ester (ACLE) could be readily incorporated into this synthesis to prepare Mtx-Cephem-Dex (Fig. 5).

Chemical Complementation: A Reaction-Independent Genetic Assay

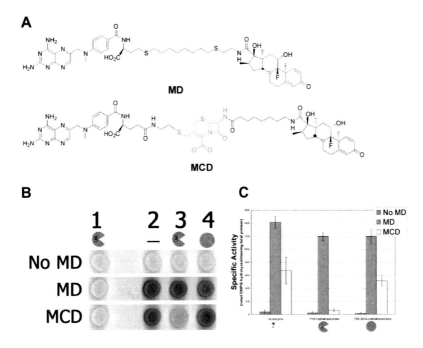

Figure 5. Chemical Complementation Links Enzyme Catalysis to Reporter Gene Transcription. **A)** Structures of the Mtx-Dex (MD) and Mtx-Cephem-Dex (MCD) heterodimers. **B)** X-gal plate assays of cephalosporinase-dependent Mtx-Cephem-Dex-induced *lacZ* transcription. Yeast strains containing a *lacZ* reporter gene were grown on X-gal indicator plates with or without Mtx-linker-Dex molecules as indicated. Columns **1-4** correspond to yeast strains containing a LexA DNA-binding domain fusion protein, a B42 activation domain fusion protein, and enzyme, as follows: **1)** LexA-DHFR, B42, P99 cephalosporinase. 1 lacks GR and is used as a negative control. **2)** LexA-DHFR, B42-GR, no enzyme; **3)** LexA-DHFR, B42-GR, P99 cephalosporinase. **4)** LexA-DHFR, B42-GR, P99 Ser64→Ala cephalosporinase. The rows correspond to individual X-gal plates, which have different small molecules as indicated: **No MD)** No Mtx-Dex; **MD)** 1 µM Mtx-Dex; **MCD)** 10 µM Mtx-Cephem-Dex. **C)** ONPG liquid assays. Yeast strains expressing the LexA-DHFR and B42-GR fusion proteins and containing a *lacZ* reporter gene and expressing either no enzyme, --, P99 cephalosporinase ![icon] ,or P99 Ser64→Ala cephalosporinase. ● , were grown in liquid culture and assayed for β-galactosidase activity using ONPG as a substrate. The liquid culture contained small molecules as indicated. The assays were done in triplicate. ONPG hydrolysis rates are reported as nmol/min/mg total protein, and the error bars for the specific activity correspond to the standard deviation from the mean. Strains containing the active P99 cephalosporinase showed an 8-fold decrease in the level of *lacZ* transcription relative to strains containing the inactive Ser64→Ala variant.

Using standard β-galactosidase activity assays both on plates and in liquid culture, we showed that Mtx-Cephem-Dex activates *lacZ* transcription in a yeast strain containing the LexA-DHFR and B42-GR fusion proteins and a *lacZ* reporter gene (Figs 5B and 5C, and data not shown). Having shown that Mtx-Cephem-Dex is an efficient dimerizer *in vivo*, we wanted to confirm that it is a good substrate for the cephalosporinase enzyme using purified enzyme *in vitro*.

The P99 cephalosporinase was subcloned into a T7 expression system with a C-terminal His6-tag, over-expressed, and purified using a Nickel-affinity resin. Since Mtx has a strong absorbance at 264 nm, turnover could not be determined based on the change in the absorbance at 264 nm upon cleavage of the lactam bond as is standard [36]. Thus, a coupled assay using 5,5'-dithiobis(2-nitrobenzoic acid) (Ellman's Reagent) was developed for measuring Mtx-Cephem-Dex hydrolysis. Upon cleavage of the β-lactam bond and expulsion of the Mtx thiol from the C3' position of the cephem, the Mtx thiol reduces Ellman's Reagent, leading to an increase in the absorbance at 412 nm. Using this coupled assay, the P99 cephalosporinase was shown to turnover Mtx-Cephem-Dex with a specific activity of 0.309 ± 0.049 µmol/min/mg of enzyme. By comparison, one of the best substrates for the *E. cloacae* P99 cephalosporinase, nitrocefin, is turned over with a specific activity of 208.7 ± 42.7 µmol/min/mg of enzyme.

Having shown that the Mtx-Cephem-Dex substrate retained the ability to activate transcription *in vivo* and was cleaved effectively by the cephalosporinase enzyme *in vitro*, the next step was to add the enzyme to the yeast three-hybrid system and find conditions where we could observe enzyme-dependent transcription. Towards this end, the yeast three-hybrid strain was re-engineered so that the expression of the transcriptional activator fusion proteins and the enzyme could be regulated independently. The LexA-DHFR and B42-GR proteins were placed under the control of the fully regulatable GAL1 promoter; and the P99 cephalosporinase, under the repressible MET promoter. First, we established independently that the cephalosporinase was being expressed in an active form in the yeast cells using nitrocefin, a known chromagenic substrate for the cephalosporinase (data not shown) [37]. Then, using standard lacZ assays on X-gal plates, we developed conditions where expression of the P99 cephalosporinase disrupted Mtx-Cephem-Dex-mediated *lacZ* transcription (Fig. 5B, column 3). These results were confirmed using quantitative assays in liquid culture with ONPG (Fig. 5C) and establish that the three-hybrid assay can be used to detect cephalosporinase activity.

A number of experiments were carried out to confirm that the change in transcription of the reporter gene was in fact due to enzyme turnover. First, we showed that enzyme-dependent disruption of the transcription read-out was cephem dependent by comparing the levels of *lacZ* transcription with Mtx-Dex and Mtx-Cephem-Dex. Using standard *lacZ* transcription assays both on plates and in liquid culture, we determined the levels of *lacZ* transcription in yeast strains expressing different LexA and B42 fusion proteins, enzymes, and a *lacZ* reporter gene (Figs 5B and 5C).

In these assays, either no small molecule, Mtx-Dex with a non-cleavable linker at 1 μM concentration in the media, or Mtx-Cephem-Dex with the cleavable cephem linker at 10 μM concentration was used. As can be seen, *lacZ* transcription in the strain expressing LexA-DHFR and B42-GR is small-molecule dependent (Figs 5B, column 2, and 5C). Expression of the wild-type cephalosporinase enzyme disrupts this small-molecule induced transcription activation when the cells are grown in the presence of Mtx-Cephem-Dex, but not Mtx-Dex (Figs 5B, column 3, and 5C). Importantly, expression of the cephalosporinase enzyme has little effect on the levels of Mtx-Dex-activated *lacZ* transcription (Figs 5B, row 2, and 5C).

Another important control is to establish that disruption of *lacZ* transcription is due to turnover of the cephem linkage and not simply sequestration of the Mtx-Cephem-Dex substrate by the cephalosporinase enzyme. To address this question, we compared the activity of the wild-type cephalosporinase enzyme in this assay with that of an inactive mutant. For the inactive variant, a mutant enzyme in which alanine replaced the active-site serine nucleophile was prepared. Cephalosporinases are serine-protease-type enzymes, and Ser^{64} is the active-site serine known to be essential for turnover of the cephem substrate [33]. In contrast with the wild-type cephalosporinase, there was no detectable change in the levels of *lacZ* transcription for cells expressing the $Ser^{64} \rightarrow$ Ala mutant enzyme (Figs 5B, columns 3 and 4, and 5C). The optimal difference in signal between the active and inactive enzyme was observed when the cells were grown in 0.5% galactose, 1.5% glucose and 134 mM methionine. Under these conditions, the three-hybrid fusion proteins are expressed at low levels, and the enzyme is maximally expressed. Together, these results establish that the change in transcription of the *lacZ* reporter gene is due to enzyme-catalysed turnover of the Mtx-Cephem-Dex substrate.

Finally, a *lacZ* screen was used to isolate the wild-type cephalosporinase from a pool of inactive variants (Fig. 6). A mixture of plasmids encoding the wild-type enzyme (5 %) or the $Ser^{64} \rightarrow$ Ala variant (95 %) was transformed en masse into a yeast strain expressing the LexA-DHFR and B42-GR fusion proteins and bearing a *lacZ* reporter plasmid. The resulting transformants were plated on X-gal plates and screened based on their levels of β-galactosidase expression. Initially, the screen suffered from a high percentage of false positives and false negatives. Integration of the genes encoding the LexA-DHFR and B42-GR fusion proteins, however, stabilized the transcription read-out without significantly affecting the sensitivity of the assay. Using the integrated strains, approximately 5% of the cells showed reduced levels of β-galactosidase expression, as would be expected based on the plasmid ratio.

The plasmids encoding the enzyme were extracted from 5 blue and 5 white colonies and sequenced. Sequencing confirmed that 5/5 of the blue cells contained the inactive $Ser^{64} \rightarrow Ala$ mutant enzyme and 4/5 of the white cells contained the wild-type cephalosporinase enzyme. A secondary *lacZ* screen with and without Mtx-Dex could rule out false positives. The *lacZ* screen demonstrates that the yeast three-hybrid system can be used reliably to screen libraries of proteins based on catalytic activity.

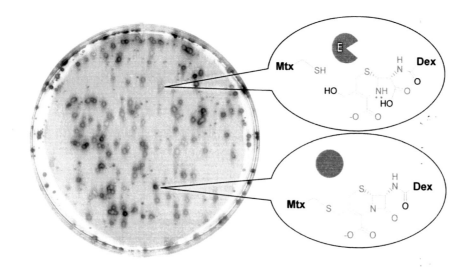

Figure 6. High-throughput Chemical Complementation Screen. Active enzyme can be isolated from a pool of inactive mutants. The yeast selection strain was transformed with a 5:95 mixture of plasmids encoding the wt active cephalosporinase enzyme and the inactive $Ser^{64} \rightarrow Ala$ cephalosporinase variant, respectively, and then plated onto an X-gal indicator plate containing 10 mM Mtx-Cephem-Dex. Cells containing the active enzyme could be distinguished based on the levels of X-gal hydrolysis and hence *lacZ* transcription.

Having established that the chemical complementation system could read-out cephalosporinase activity, we next evaluated the ability of the system to distinguish enzymes based on their levels of catalytic activity [38]. A series of P99 cephalosporinase variants were designed that spanned several orders of magnitude in their catalytic activity. These variants were over-expressed and purified, and then their activity with the Mtx-Cephem-Dex substrate was determined *in vitro* using Michaelis-Menten kinetics. Next these same variants were introduced into the chemical complementation system, and their levels of transcription activation were determined using standard assays for *lacZ* transcription in liquid culture.

For all of the variants tested, the catalytic efficiency (k_{cat}/K_M) of the variant correlated with its level of *lacZ* activation in the three-hybrid assay (Fig. 7). The more active variants showed decreased levels of *lacZ* transcription, presumably because they were able to cleave the Mtx-Cephem-Dex substrate efficiently. While the less active variants showed higher levels of *lacZ* transcription. Already just using these initial conditions and a single reporter gene, the assay was able to distinguish variants spanning almost three-orders of magnitude in k_{cat}/K_M. The ability to distinguish enzymes based on their catalytic activity will be crucial for directed evolution experiments where the goal is to evolve variants with increased activity.

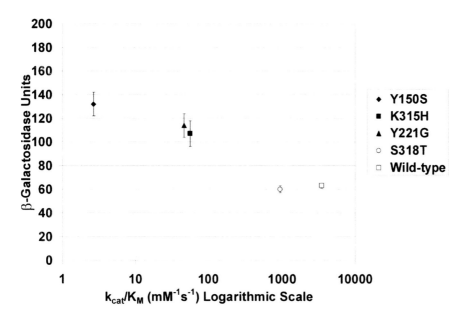

Figure 7. Correlation between catalytic efficiency and in vivo transcription of the wild-type P99 β-lactamase and its variants. Errors in k_{cat}/K_m were negligible and not plotted. Standard deviations for *lacZ* transcription activation are shown. The standard deviation for the wild-type P99 is 2 units, and is subsumed by the symbol on the graph.

Finally, in unpublished results, we have used the chemical complementation system for the directed evolution of an enzyme that catalyses the formation of a glycosidic bond (Lin, Tao and Cornish, V. unpublished results). Significantly, this result shows the generality of the approach and the ease with which it can be applied to new chemical reactions. This result also demonstrates that chemical complementation can detect bond formation as well as bond cleavage reactions and that enzyme catalysis can be linked to a growth selection as well as a *lacZ* screen.

CONCLUSION

While demonstrated with the cephalosporinase enzyme, the advantage of this selection strategy should be its generality. The yeast three-hybrid system should be able to link both bond cleavage and bond formation reactions to transcription of a reporter gene. In the case of bond cleavage reactions, the enzyme should ensure cell survival by cleaving a Mtx-Dex substrate and disrupting transcription of a toxic reporter gene. For bond formation, the enzyme should form a bond between Mtx and Dex, activating transcription of an essential reporter gene. The read-out system - Mtx-Dex, LexA-DHFR, B42-GR, and the reporter gene - can all remain constant while the chemistry changes. Thus, all that needs to be changed for each new reaction is the Mtx-SUBSTRATE-Dex or Dex-SUBSTRATE and Mtx-SUBSTRATE molecules synthesized in the lab and the enzyme library. This assay could be used to engineer glycosyltransferases, aldolases, esterases, amidases, and Diels-Alderases - all with a variety of substrate specificities and regio- and stereo-selectivities. By converting the assay to a coupled enzyme assay, it may even be possible to detect oxidases and reductases. In addition to providing a powerful selection for the evolution of enzymes with new activities, there should be many uses for a reaction-independent, high-throughput assay for enzyme catalysis. The assay can be used to study enzyme function, either to test hundreds of mutants to identify amino acids important for the catalytic activity of an enzyme or hundreds of different molecules to determine the substrate specificity of an enzyme. Likewise, the assay could be applied to drug discovery by screening libraries of small molecules based on inhibition of enzyme activity and a change in transcription of the reporter gene. A distinct advantage is that a new assay would not have to be developed for each new enzyme target. Finally, this assay should be particularly well suited to proteomics. A battery of Mtx-Dex substrates with different substrates as linkers could be prepared and then used to screen cDNA libraries to identify enzymes that fall into common families, such as glycosidases or aldolases. Since mammalian, as well as yeast, three-hybrid assays are standard, the assays could be carried out in the endogenous cell line ensuring correct post-translational modification of the proteins. The key to all of these applications is a robust assay for enzymatic activity.

ACKNOWLEDGEMENTS

We are grateful for financial support for this work from the Burroughs-Wellcome Foundation, the Beckman Foundation, and the National Institutes of Health (RO1-GM062867). VWC is a recipient of a Beckman Young Investigator Award, a Burroughs Wellcome Fund New Investigator Award in the Toxicological Sciences, a Camille and Henry Dreyfus New Faculty Award, a NSF CAREER Award, and a Sloan Foundation Fellowship. We thank Otsuka Chemical Co. for kindly providing the advanced cephem intermediate ACLE. VWC thanks R. Sauer for allowing her to initiate this project as a post-doctoral fellow in his laboratory.

REFERENCES

[1]   Lin, H., Cornish, V. (2002) *Angew. Chem. Int. Ed.* **41**:4402-4425.

[2]   Arnold, F. (2001) *Nature* **409**:253-257.

[3]   Yano, T., Oue, S., Kagamiyama, H. (1998) *Proc. Natl. Acad. Sci. USA* **95**: 5511-5515.

[4]   Joo, H., Lin, Z., Arnold, F. (1999) *Nature* **399**:670-673.

[6]   Baum, E., Bebernitz, G., Gluzman, Y. (1990) *Proc. Natl. Acad. Sci. USA* **87**:10023-10027.

[7]   Smith, T., Kohorn, B. (1991) *Proc. Natl. Acad. Sci. USA* **88**:5159-5162.

[8]   Hermes, J., Blacklow, S., Knowles, J. (1990) *Proc. Natl. Acad. Sci. USA* **87**:696-700.

[9]   Kast, P., Asif-Ullah, M., Jiang, N., Hilvert, D. (1996) *Proc. Natl. Acad. Sci. USA* **93**: 5043-5048.

[10]   Schultz, P., Lerner, R. (1995) *Science* **269**:1835-1842.

[11]   Koltermann, A., Kettling, U., Bieschke, J., Winkler, T., Eigen, M. (1998) *Proc. Natl. Acad. Sci. USA* **95**:1421-1426.

[12]   Olsen, M., Stephens, D., Griffiths, D., Daugherty, P., Georgiou, G., Iverson, B. (2000) *Nature Biotechnol.* **18**:1071-1074.

[13]   Pedersen, H., Holder, S., Sutherlin, D., Schwitter, U., King, D., Schultz, P. (1998) *Proc. Natl. Acad. Sci. USA* **95**:10523-10528.

[14]   Atwell, S., Wells, J. (1999) *Proc. Natl. Acad. Sci. USA* **96**:9497-9502.

[15]   Firestine, S., Salinas, F., Nixon, A., Baker, S., Benkovic, S. (2000) *Nature Biotechnol.* **18**:544-547.

[16]   Lin, H., Abida, W., Sauer, R., Cornish, V. (2000) *J. Am. Chem. Soc.* **122**:4247-4248.

[17]   Licitra, E., Liu J. (1996) *Proc. Natl. Acad. Sci. USA* **93**:12817-12821.

[18] Spencer, D., Wandless, T., Schreiber, S., Crabtree, G. (1993) *Science* **262**:1019-1024.

[19] Clackson, T. et al. (1998) *Proc. Natl. Acad. Sci. USA* **95**:10437-10442.

[20] Chakraborti, P., Garabedian, M., Yamamoto, K., Simons, S. (1991) *J. Biol. Chem.* **266**:22075-22078.

[21] Picard, D., Yamamoto, K. (1987) *EMBO J.* **6**:3333-3337.

[22] Govindan, M., Manz, B. (1980) *Eur. J. Biochem.* **108**:47-53.

[23] Manz, B., Heubner, A., Kohler, I., Grill, H.-J., Pollow, K. (1983) *Eur. J. Biochem.* **131**:333-338.

[24] Bolin, J., Filman, D., Matthews, D., Hamlin, R., Kraut, J. (1982) *J. Biol. Chem.* **257**:13663-13672.

[25] Benkovic, S., Fierke, C., Naylor, A. (1988) *Science* **239**:1105-1109.

[26] Sasso, S., Gilli, R., Sari, J., Rimet, O., Briand, C. (1994) *Biochim. Biophys. Acta* **1207**:74-79.

[27] Hart, B.P., Haile, W.H., Licato, N.J., Bolanowska, W.E., McGuire, J.J., Coward, J.K. (1996) *J. Med. Chem.* **39**:56-65.

[28] Fields, S., Song, O. (1989) *Nature* **340**:245-246.

[29] Gyuris, J., Golemis, E., Chertkov, H., Brent, R. (1993) *Cell* **75**:791-803.

[30] Ausubel, F., Brent, R., Kingston, R., Moore, D., Seidman, J., Smith, J., Struhl, K. (1995) *Current Protocols in Molecular Biology.* John Wiley & Sons, New York.

[31] Adams, A., Gottschling, D., Kaiser, C., Stearns, T. (1998) *Methods in Yeast Genetics.* Cold Spring Laboratory Press, Plainview.

[32] Baker, D., Bleckzinski, C, Lin, H., Salazar-Jimenez, G., Sengupta, D., Krane, S., Cornish, V.W. (2002) *Proc. Natl. Acad. Sci. USA* **99**:16537-16542.

[33] Galleni, M., Frere, J. (1988) *Biochem. J.* **255**:119-122.

[34] Lobkovsky, E. et al. (1993) *Proc. Natl. Acad. Sci. USA* **90**:11257-11261.

[35] Durckheimer, W., Adam, F., Fischer, G., Kirrstetter, R. (1988) *Adv. Drug Res.* **17**:61-234.

[36] Page, M. (1992) *The Chemistry of the β-Lactams.* Chapman & Hall, Glasgow.

[37] Pluckthun, A., Knowles, J. (1987) *J. Biol. Chem.* **262**:3951-3957.

[38] Sengupta, D., Muller, R., Cornish, V.W. (2004) *Biochemistry* **22**:5767-5768.

[39] Abida, W., Carter, B., Althoff, E., Lin, H., Cornish, V.W. (2002) *ChemBiochem.* **3**:887-895.

The Chemical Theatre of Biological Systems, May 24th - 28th, 2004, Bozen, Italy

VIRTUAL SCREENING USING BINARY KERNEL DISCRIMINATION

JÉRÔME HERT, PETER WILLETT AND DAVID J. WILTON

Krebs Institute for Biomolecular Research and Department of Information Studies, University of Sheffield, Western Bank, Sheffield S10 2TN, U.K.

E-Mail: *p.willett@sheffield.ac.uk

Received: 7th June 2004 / Published: 22nd July 2005

ABSTRACT

This paper discusses a range of procedures for virtual screening of chemical databases in which the molecules are represented by 2D fragment bit-strings. Training-sets containing active and inactive molecules from the NCI AIDS dataset and from the Syngenta corporate pesticide file were processed using binary kernel discrimination (BKD), similarity searching, and substructural analysis methods. The effectiveness of these methods was judged by the extent to which active test-set molecules were clustered towards the top of the resultant rankings. The BKD approach was found consistently to yield the best rankings, and its general effectiveness was confirmed in similarity searches of the MDL Drug Data Report database based on multiple reference structures. As well as being effective, BKD is reasonably efficient, and the method would hence appear to be well suited to virtual screening of 2D structure databases.

INTRODUCTION

Increasing use is being made of virtual screening methods to maximize the cost effectiveness of biological screening programmes by prioritizing a test-set, such as a corporate database of previously unassayed molecules, in decreasing order of the *a priori* probability of activity [1, 2].

There are many different types of virtual screening method: in the first part of this paper, we discuss machine learning methods that can be used when heterogeneous sets of both active and inactive molecules are available for use as training data, e.g., after an initial round of high-throughput screening (HTS).

Specifically, we report a comparison of several such methods when they are used in simulated virtual screening experiments with the NCI AIDS database [3] and with pesticide data from the Syngenta corporate file. These experiments, based on 2D fragment bit-strings, demonstrate the effectiveness of the machine learning technique known as *binary kernel discrimination* (BKD) [4].

BKD is studied further in the second part of the paper, which considers the use of similarity searching as a virtual screening mechanism. Similarity searching is normally used to identify those database molecules that are most similar to a single active *reference structure*, using some quantitative definition of inter-molecular structural similarity [5,6]. Following Schuffenhauer *et al.* [7] and Xue *et al.* [8], we consider here similarity-based virtual screening techniques that can be used when not one but several different bioactive reference structures are available, i.e., when the training-set contains only active molecules. A comparison of several search methods that make use of multiple reference structures, this time using the MDL Drug Data Report (MDDR) database [9], further demonstrates the general effectiveness of BKD for virtual screening of 2D chemical structure databases. Many additional studies, including the use of other structure representations and other screening methods, are reported by Wilton *et al.* [10] and by Hert *et al.* [11].

VIRTUAL SCREENING WITH ACTIVE AND INACTIVE TRAINING DATA

Virtual Screening Methods

Similarity methods

The simplest way of predicting the likely activities of a set of molecules is by computing their similarities to a training-set of known actives and inactives, i.e., a *k*-nearest neighbour classifier. Two approaches are reported here; experiments with other, less effective similarity approaches are discussed by Wilton *et al.* [10]. Given a test-set molecule, *j*, to be predicted, S_{max} is defined to be the similarity to the most similar training-set active, i.e.,

$$S_{max}(j) = \max\{S(i, j)\} \quad i \in Actives$$

while *SA-I* is defined to be the mean similarity to all the training-set actives minus the mean similarity to all the training-set inactives, i.e.,

$$S_{A-I} = \frac{1}{N_A} \sum_{i \in Actives} S(i, j) - \frac{1}{N_I} \sum_{i \in Inactives} S(i, j)$$

The similarities in these approaches are calculated using the Tanimoto Coefficient [5]. If a, b and c are the numbers of bits set in the one fingerprint, set in the other fingerprint, and set in both fingerprints, respectively, then the Tanimoto coefficient is defined to be:

$$\frac{c}{a + b \angle c}$$

In these methods, as with all the ohers considered here, the test-set molecules are ranked in descending order of the calculated scores, i.e. similarity values in the present context, with the expectation that the top-ranked molecules have the greatest probability of activity.

Substructural analysis methods

Substructural analysis was first described by Cramer *et al.* [12], and many different weighting schemes have been described for this purpose [13]. For each fragment or bit, j, in the binary fingerprints that characterize the training-set molecules, a weight is calculated that is a function of the numbers of active and inactive molecules in the training set that have the j-th bit set. A score is then computed for a test-set molecule by summing (or otherwise combining) the weights of those bits that are set in its fingerprint. Here, we report the use of the R1 and R2 weights that performed well in the comparative study of Ormerod *et al.* [13]. Let A_j (I_j) be the number of active (inactive) molecules with bit j set, and let T_j be the total number of molecules with bit j set. Similarly, let N_A (N_I) be the total number of active (inactive) molecules, and let N_T be the total number of molecules in the training-set. Then the two weights are given by:

$$R1 = \log\left(\frac{A_j / N_A}{T_j / N_T}\right), \quad R2 = \log\left(\frac{A_j / N_A}{I_j / N_I}\right).$$

Binary kernel discrimination

The use of binary kernel discrimination (BKD) for chemical applications has been described by Harper *et al.* [4]. For two molecules i and j, characterized by binary fingerprints of length M that differ in d_{ij} positions, they suggest the use of the kernel function K_λ

$$K_\lambda(i,j) = \lambda^{M-d_{ij}}(1-\lambda)^{d_{ij}}$$

where λ is a smoothing parameter the value of which is to be determined. Training-set molecules are then ranked using the scoring function:

$$L_A(j) = \frac{\sum\limits_{i\in Active} K_\lambda(i,j)}{\sum\limits_{i\in Inactive} K_\lambda(i,j)}$$

with the optimum value of λ being found from analysis of the training-set. The optimum is obtained by computing scores for each training-set molecule using the other training-set molecules for a number of different values of λ in the range 0.50 to 0.99. For each value of λ the sum of the ranks of the active molecules is computed. If this is plotted against λ a clear minimum should be observed indicating the optimum λ, i.e. the value that minimizes the summed ranks of the actives in the training-set. It is assumed that the optimal value in the training-set is also optimal for the test-set. This is clearly a strong assumption, but the results we have obtained suggest that it does not result in poor predictive performance and it is difficult to use a machine-learning technique such as this without an assumption.

Results and Discussion

We carried out simulated virtual screening experiments on two datasets, one public and one corporate. The initial experiments used the NCI AIDS file [3], which contains molecules that have been checked for anti-HIV activity, with 1129 confirmed actives or confirmed moderately actives and 34,862 inactives. Training-sets were randomly generated, each containing 200 actives and 200 inactives, with the remaining 35,591 molecules forming the test-set: three such training-sets were generated for the experiments.

The Syngenta dataset contained 132,784 molecules that had been tested in various in vivo whole organism screens; of these 7127 were active in at least one screen, with the remaining 125,657 having a response in the screens less than a pre-defined threshold value. As before, three different training-sets were randomly generated, each containing 713 actives (i.e., 10% of the total actives) and 713 inactives with the remaining 131,358 molecules in each case forming the test-sets.

The test-set and training-set molecules were represented by 988-bit Tripos Unity fingerprints [14] and scores were calculated for each of the test-set molecules. The test-set was then ranked, and the effectiveness of the various methods determined by noting the numbers of actives in the top-1% and the top-5% of the ranking. The results obtained with the three different training-sets were all very similar, in that though there were variations in the precise values obtained with the different test-sets there was very little difference in the relative performance of the various methods; we hence consider the results for only one of the training-sets, as detailied in Table 1.

Table 1. Percentages of the active molecules retrieved in the top 1% and in the top 5% of the rankings for the NCI and Syngenta datasets.

Ranking Method	NCI Dataset		Syngenta Dataset	
	Top-1%	Top-5%	Top-1%	Top-5%
S_{max}	8.18	38.86	5.85	19.05
S_{A-I}	10.44	21.42	6.61	19.10
$R1$	11.52	21.42	6.81	18.69
$R2$	1.51	16.47	6.93	20.55
BKD	13.67	42.73	9.84	27.14

Inspection of this table leads to a simple, unequivocal conclusion: that the BKD method gives rankings that are far superior to those of the other ranking methods considered here, and thus that this is the method of choice in terms of the effectiveness of virtual screening. The difference between BKD and the other approaches in the table (and the many other approaches discussed by Wilton *et al.* [10]) is particularly marked with the large Syngenta dataset.

A highly effective virtual screening method is of little practical use if it cannot be applied to datasets of realistic size. The computational requirements are in three parts: the analysis of the training-set, which in BKD requires repeated processing of the training-set to identify the optimal value of the parameter λ; the scoring of each of the molecules in the test-set using the $L_A(j)$ scoring scheme defined previously; and then the ranking of the test-set in descending order of the calculated scores. Even so, the method's computational requirements are not overly large. Using programs written in C and run on a Silicon Graphics R12000 processor, training for the very large Syngenta dataset took about 36 seconds for each value of l that was tested, and the subsequent scoring using the optimal value of l took 3730 CPU seconds.

VIRTUAL SCREENING WITH ACTIVE TRAINING DATA

The increasing use of HTS means that large amounts of active and inactive training data are likely to become available shortly after the commencement of an agrochemical or pharmaceutical lead-discovery programme. Right at the start of a programme, however, the medicinal chemist may have knowledge of just a few active molecules, such as natural products, or patented, competitor molecules. In such cases, an alternative type of virtual screening can be used, based on similarity searching [5, 6]. This involves matching a single bioactive reference structure against each of the database structures to identify those that are most similar (and hence most likely to exhibit the same activity as the reference structure); here, we evaluate three distinct approaches that can be used when multiple reference structures are available, i.e., when the training-set contains only actives.

Virtual Screening Methods
Modal fingerprint method

Shemetulskis *et al.* first described the single fingerprint approach in their work on Stigmata [15]. The method generates a modal fingerprint from an input set of molecules that seeks to capture the common chemical features present in the members of this training-set. A bit j is set to "on" in the modal fingerprint if that bit is found in more than a user-defined threshold percentage of the training-set molecules.

The modal fingerprint is then used as a query and compared to the fingerprints of the molecules in the test-set. Shemetulskis *et al.* used two metrics to rank the molecules of the database: the modal percent and the Tanimoto coefficient, and we have used the latter approach here.

Data fusion method

Data fusion is the name given to a range of techniques that combine inputs from different sensors, with the expectation that using multiple information sources enables more effective decisions to be made than if just a single sensor is employed. The approach has been made in many different fields; when applied to chemoinformatics applications (where it is sometimes referred to as consensus scoring) the fusion is effected by combining the results of several database searches using different descriptors or scoring functions [16]. In conventional applications of data fusion, a single active reference structure is searched against a database in several different ways; in the present context, we have several different reference structures that are all searched against a database in exactly the same way (specifically using 2D fingerprints with the Tanimoto coefficient). We have fused the similarity scores, $S(i,j)$ ($1 \leq i \leq n$, the number of active reference structures) for some molecule j using the S_{max} fusion rule discussed in the first section of the paper, i.e., the test-set molecules were ranked on the basis of :

$$\max \{ S(i, j) \} \quad i \in Actives$$

Substructural analysis method

A weighting scheme for substructural analysis normally requires access to training-set data for both actives and inactives. In the present context, however, we do not have access to all the necessary information as the training-set consists of just active molecules. However, if we restrict our attention to those weighting schemes that do not make explicit use of information about the inactives and also make the assumption that the overall characteristics of the training-set are mirrored by those of the entire database that is to be searched, then we can use the R1 weight. The equation for the weight is as given previously; here, however, T_j is the total number of molecules in the database with the bit j set and N_T is the total number of molecules in the database (rather than the total numbers of molecules in the training-set, as in conventional substructural analysis).

Binary kernel discrimination

An analogous approximation can be used to enable BKD to be used when only actives are available for training purposes. The approach we have taken is to make the assumption that the overall characteristics of the inactives are approximated with a high degree of accuracy by the overall characteristics of the entire database that is to be searched. If this assumption is accepted then a training-set can be generated by taking the set of reference structures and adding to it molecules randomly selected from the database, with the expectation that most, if not all, of these added molecules are inactive. Since actives are inherently very rare, both this expectation and that underlying the R1 approximation are not unreasonable.

Results and Discussion

We have evaluated the various approaches above by means of simulated virtual screening searches on the MDDR database [9]. After removal of the duplicates and molecules that could not be processed using local software, a total of 102,535 molecules was available for searching and these were represented by Unity fingerprints [14]. These molecules were searched using the eleven sets of active molecules from MDDR that are listed in Table 2. A rough guide to the diversity of each of the chosen sets of bioactives is provided by matching each compound with every other in its activity class, calculating similarities using the Unity fingerprint and Tanimoto coefficient and computing the mean of these intra-set similarities. The resulting similarity scores are listed in the second column of Table 2, where it will be seen that the renin inhibitors are the most homogeneous and the cyclooxygenase inhibitors are the most heterogeneous.

For each of the 11 activity classes, ten active molecules were selected for use as the training-set. The selections were done at random, subject to the constraint that no pair-wise similarity in a group exceeded 0.80 (using Unity fingerprints and the Tanimoto coefficient). Each searching method was repeated ten times using different training-sets, and in each search, a note was made of the percentage of the active molecules (i.e., those in the same class as those in the training-set) that occurred in the top 5% of the ranking resulting from that search. The results presented below are the mean and standard deviations for these recall values, averaged over each set of ten searches (very similar results to those listed here were obtained if the top-1%, rather than the top-5%, of the rankings were evaluated).

Table 2. Mean percentage of active molecules retrieved by multiple-reference methods over the top 5% of the ranked test-set.

Activity Class	Self-Similarity	Modal	Data Fusion	Substructural Analysis	BKD
5HT3 antagonist	0.35	30.31	49.03	29.27	52.32
5HT1A agonist	0.34	21.85	37.15	30.13	38.19
5HT Reuptake inhibitor	0.34	39.63	49.68	33.12	45.82
D2 antagonist	0.34	27.12	37.40	27.51	38.65
Renin inhibitor	0.57	88.77	88.62	52.94	93.34
Angiotensin II AT1 antagonist	0.40	73.63	80.44	43.40	84.47
Thrombin inhibitor	0.41	49.43	58.58	35.64	63.06
Substance P antagonist	0.39	36.80	47.14	36.52	58.39
HIV protease inhibitor	0.44	53.53	61.62	34.05	68.45
Cyclooxygenase inhibitor	0.26	10.96	26.52	19.20	33.15
Protein kinase C inhibitor	0.32	35.60	48.01	35.58	49.37
Average over all classes	0.38	42.51	53.11	34.31	56.84

A large number of searches was carried out to identify the best parameter settings for the various methods discussed above (and also several other, less effective methods that are described by Hert *et al.* [11]). The resulting settings were then used in the main experiments, the results of which are detailed in the body of Table 2. Inspection of this table shows that the fusion of the similarity scores and BKD are the clear methods of choice, consistently out-performing modal fingerprints and substructural analysis.

With some minor exceptions, the performance of all of the methods tends to increase as the self-similarity of the active molecules increases. The correlation with intra-class similarity is not unexpected; what is of importance is that good screening performance is obtained even with quite diverse activity classes (such as the protein kinase C inhibitors and the D2 antagonists). The worst results are obtained with the most diverse set of actives, i.e., the cyclooxygenase inhibitors; even here, however, the data fusion and BKD runs represent 5.3-fold and 6.6-fold enrichments, respectively, over a random ranking of the dataset (with average enrichment factors for these two approaches over all classes being 10.6 and 11.3, respectively).

In the final set of experiments, we sought to quantify the benefit that can be achieved using multiple reference structures, rather than single reference structures as in conventional similarity searching. This was done by using every single active molecule in turn in each of the 11 chosen activity classes as the reference structure, and recording the minimum, mean and maximum performance, as detailed in Table 3.

Table 3. Percentage of active molecules retrieved by single similarity searches over the top 5% of the ranked test-set.

Activity Class	Mean	Max	Min
5HT3 antagonist	21.15	40.97	1.89
5HT1A agonist	18.43	39.29	2.45
5HT Reuptake inhibitor	24.02	42.69	1.43
D2 antagonist	17.35	35.58	0.26
Renin inhibitor	80.54	93.21	2.95
Angiotensin II AT1 antagonist	48.04	81.67	3.64
Thrombin inhibitor	33.51	63.56	0.63
Substance P antagonist	26.87	57.69	0.57
HIV protease inhibitor	37.60	63.65	1.89
Cyclooxygenase inhibitor	9.39	21.09	0.32
Protein kinase C inhibitor	19.42	46.05	0.68
Average over all classes	30.57	53.22	1.52

The mean values correspond to the performance that might be expected using an average, individual reference structure and are clearly much lower than the figures reported in Table 2 for the BKD and data fusion methods (30.57% as against 56.84% and 53.11%, respectively). Thus, the use of ten actives, rather than just one, results in an increase of over two-thirds in the numbers of actives retrieved. Perhaps the most interesting figures in Table 3 are those listed under "Max". These represent the best single similarity searches possible from the many hundreds of individual bioactive molecules (this number ranges from 349 for the 5HT re-uptake inhibitors up to 1236 for the substance P antagonists). If we consider the average over all activity classes, it will be seen that this upper-bound is only fractionally better than the data fusion result in Table 2 and is actually worse than the BKD figure. Thus, on average, picking any ten active reference structures and combining them using S_{max} data fusion or BKD will enable searches to be carried out that are comparable to even the best possible conventional similarity search using a single active reference structure.

Choosing between S_{max} data fusion and BKD is difficult; the latter is more effective here, but is much more time-consuming; however, the first set of experiments show that it is far more effective than data fusion when large amounts of training-data are available, rather than the very small sets of ten actives used here. We hence conclude that BKD will, in general, result in better rankings of a test-set than will the far simpler data fusion approach.

CONCLUSIONS

Virtual screening provides an ideal domain for the application of machine learning techniques, many of which are designed specifically for binary categorization problems (the categories here being active and inactive). At the same time, the characteristics of chemical datatsets are very different from those common in much machine-learning research: very extensive use is made of binary, rather than real-valued, object representations (i.e., fragment bit-strings); the data are very numerous (chemical datasets are typically of size 10^5-10^6); and the two categories are markedly different in size (since actives are far, far less common than inactives). It is thus of some interest to test new machine-learning methods in this domain, and this has been the principal driver for our studies of the use of kernel discrimination methods for virtual screening. It is very important not to over-emphasize the advantages of some new computational approach. Even so, the results presented here do suggest that BKD provides an attractive focus for future research: it is effective in operation, in both exact and approximate forms; it has been applied successfully to both pharmaceutical and agrochemical datasets; and it can be used with very large files of 2D fingerprints.

We can regard a kernel function as a new type of similarity measure and it is hence of interest to consider the three components of a chemical similarity measure [5], *viz* the structure representation, the similarity coefficient and the weighting scheme that are used. We intend to investigate all of these aspects in the context of kernel functions. For example, previous work has shown that the Hamming Distance is far less effective for fingerprint-based similarity searching than coefficients such as the Tanimoto Coefficient and the Cosine Coefficient, and it would therefore be of interest to consider the use of these, and other, coefficients with other types of kernel function. Again, one could employ weighted fingerprints, with the bits being assigned weights derived from substructural analysis of training-set molecules, or look at alternative types of representation. We intend to investigate these developments of the basic BKD approach in the future.

ACKNOWLEDGEMENTS

We thank the following: the Novartis Institutes for Biomedical Research for funding J.H.; Syngenta for funding D.J.W.; John Delaney, Kevin Lawson and Graham Mullier (Syngenta) and Pierre Acklin, Kamal Azzaoui Edgar Jacoby and Ansgar Schuffenhauer (Novartis) for helpful comments on this work; MDL Information Systems Inc. for the provision of the MDDR database; and Barnard Chemical Information Ltd., Daylight Chemical Information Systems Inc., the Royal Society, Tripos Inc. and the Wolfson Foundation for software and laboratory support. The Krebs Institute for Biomolecular Research is a designated biomolecular sciences centre of the Biotechnology and Biological Sciences Research Council.

REFERENCES

[1]   Böhm, H.-J., Schneider, G. (Eds) (2000) *Virtual Screening for Bioactive Molecules.* Wiley-VCH, Weinheim.

[2]   Klebe, G. (Ed.) (2000) *Virtual Screening: an Alternative or Complement to High Throughput Screening.* Kluwer, Dordrecht.

[3]   The NCI AIDS database is available at URL http://dtp.nci.nih.gov/. The details of the NCI assay are at URL http://dtp.nci.nih.gov/docs/aids/anti-hiv-screening.html

[4]   Harper, G., Bradshaw, J., Gittins, J.C., Green, D.V.S., Leach, A.R. (2001) Prediction of biological activity for high-throughput screening using binary kernel discrimination. *J. Chem. Inf. Comput. Sci.* **41**:1295-1300.

[5]   Willett, P., Barnard, J.M., Downs, G.M. (1998) Chemical similarity searching. *J. Chem. Inf. Comput. Sci.* **38**:983-996.

[6]   Sheridan, R.P., Kearsley, S.K. (2002) Why do we need so many chemical similarity search methods? *Drug Discov. Today* **7**:903-911.

[7]   Xue, L., Stahura, F.L., Godden, J.W., Bajorath, J. (2001) Fingerprint scaling increases the probability of identifying molecules with similar activity in virtual screening calculations. *J. Chem. Inf. Comput. Sci.* **41**:746-753.

[8]   Schuffenhauer, A., Floersheim, P., Acklin, P., Jacoby, E. (2003) Similarity metrics for ligands reflecting the similarity of the target proteins. *J. Chem. Inf. Comput. Sci.* **43**:391-405.

[9]   The MDL Drug Data Report database is available from MDL Information Systems Inc. at http://www.mdli.com

[10]   Wilton, D.J., Willett, P., Lawson, K., Mullier, G. (2003) Comparison of ranking methods for virtual screening in lead-discovery programs. *J. Chem. Inf. Comput. Sci.* **43**:469-474.

[11]   Hert, J., Willett, P., Wilton, D.J., Acklin, P., Azzaoui, K., Jacoby, E., Schuffenhauer, A. (2004) Comparison of fingerprint-based methods for virtual screening using multiple bioactive reference structures. *J. Chem. Inf. Comput. Sci.* **44**:1177-1185.

[12]   Cramer, R.D., Redl, G., Berkoff, C.E. (1974) Substructural analysis. A novel approach to the problem of drug design. *J. Med. Chem.* **17**:533-535.

[13]   Ormerod, A., Willett, P., Bawden, D. (1989) Comparison of fragment weighting schemes for substructural analysis. *Quant. Struct.-Activ. Relat.* **8**:115-129.

[14]   The Unity software is available from Tripos Inc. at http://www.tripos.com

[15]   Shemetulskis, N.E., Weininger, D., Blankley, C.J., Yang, J.J., Humblet, C. (1996) Stigmata: an algorithm to determine structural commonalities in diverse datasets. *J. Chem. Inf. Comput. Sci.* **36**:862-871.

[16]   Ginn, C.M.R., Willett, P., Bradshaw, J. (2000) Combination of molecular similarity measures using data fusion. *Perspect. Drug Discov. Design* **20**: 1-16.

The Chemical Theatre of Biological Systems, May 24th - 28th, 2004, Bozen, Italy

GPCR HIT DISCOVERY BEYOND HTS

KONRAD H. BLEICHER[*,1], ALEXANDER I. ALANINE[1], MARK ROGER-EVANS[1] AND GISBERT SCHNEIDER[2]

[1]F. Hoffmann-La Roche AG, Pharmaceuticals Devision, CH 4070 Basel, Switzerland

[2]Johann Wolfgang Goethe University Frankfurt, Marie-Curie-Str. 11,
D-60439 Frankfurt, Germany

E-Mail: *konrad.bleicher@roche.com

Received: 20th September 2004 / Published: 22nd July 2005

ABSTRACT

High-throughput screening is meanwhile well established in most pharmaceutical companies. Although it is routinely applied for most biological targets, several limitations ask for alternative methodologies. This article will describe two different approaches where highly potent ligands for G-protein coupled receptor targets were identified without the application of random high-throughput screening.

INTRODUCTION

A recent analysis of the targets where drugs have been successfully developed shows that about 45% of these targets belong to receptors of which most are G-protein coupled (GPCRs)[1]. Such proteins are involved in various disease pathways and are recognized as highly valuable targets for most therapeutic indications [2,3]. Besides the obvious commercial aspects, the large number of launched drugs for G-protein coupled receptors also gives confidence in the 'drugability' of such seven transmembrane proteins. Due to their size and high lipophilicity the isolation and crystallization of GPCRs turned out to be extremely difficult, making a structure based drug design approach currently unrealistic. Therefore GPCR programmes are usually initiated after a successful high-throughput screening campaign (HTS) has been accomplished.

Chemists clearly appreciate the luxury of hit lists and compound clusters identified via HTS. Nevertheless tedious downstream work is necessary, such as hit confirmation by re-ordering/ synthesis of compounds, quality assessment as well as structural confirmation to name but a few. It is the task of lead generation chemists to deal with such issues in order to generate as much information as possible from an HTS campaign before the actual hit-to-lead programme is started. Although high-throughput assays can generally be set up independently from the target, such screening exercises are particularly interesting for proteins where no biostructural- or patent data is available. Due to increasing compound inventories, constantly upcoming novel targets and requested selectivity screening, prioritization has to take place as to where and when to initiate a high-throughput screening campaign. It is obvious that there is a constant need for complementary technologies, which allow the initiation of chemistry programmes independent from the high-throughput screening route.

TARGETED LIBRARIES FROM PRIVILEGED STRUCTURES

One such alternative source to generate novel hits is the biased testing of targeted libraries. Such small compound arrays (100-1000 cpds) can be tested manually using e.g., radio-ligand binding assays which are usually available long before an HTS compatible assay is established. Clearly, such biased libraries have to show far higher hit rates to be competitive to a random screening approach where usually an average hit rate of 0.1-1% is observed. Besides the much smaller overhead and the fact that chemistry can be initiated earlier (or even without any HTS) the quality and tractability of the compounds within a targeted library subset delivers a big advantage over historical collections where often undesired compounds such as reactive intermediates, poorly soluble compounds or chemical entities which are synthetically challenging, are discovered. The design of such compound collections is essential in this regard, not only to deliver series with a certain target bias, but also showing favourable DMPK and physico-chemical property profiles. In the area of G-protein coupled receptors, where poor biostructural information calls for alternative approaches, several successful projects have been disclosed [4].

From various strategies the 'privileged structures' approach is probably most often applied. It is still unclear which molecular features are required for a chemotype to result in a 'privileged structure' and how the target proteins recognize such chemotypes.

It has been discussed that common binding sites [5] and first contact motives at the surface of the proteins [6] might be responsible for this promiscuous binding behaviour, but a clear understanding on the binding event of 'privileged structures' to their proteins is still lacking.

The term 'privileged structures' was first coined by Evans *et al.* where benzodiazepines were described to show an inherent affinity to G-protein coupled receptors [7]. This compound class has been exhaustively explored in the GPCR area where modifications such as benzodiazepinones and -diones as well their regioisomers such as the 1,4- and 1,5-benzodiazepines (-ones etc.) and their aza-analogues have been employed.

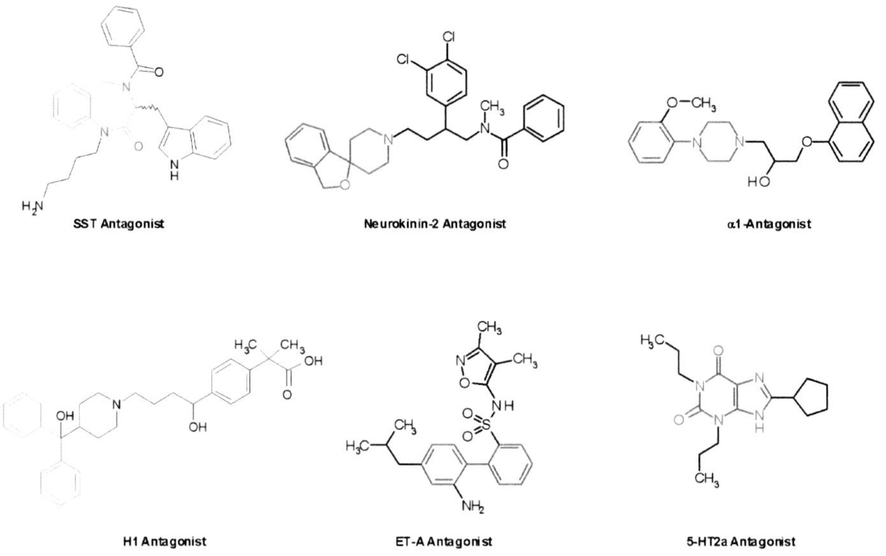

Figure 1. 'Privileged GPCR structures' (highlighted in red) as promiscuous elements of protein ligands (e.g. Benzodiazepinones, Spiropiperidines, Arylpiperazines, Biphenyl methyl-, Biphenyl-Purines).

A number of different 'privileged structures' have meanwhile been discussed in the literature, some of them are depicted in Fig. 1. Exemplified are six different chemotypes representing ligands for six different GPCRs. The spiropiperidines (e.g. represented by the Neurokinin-2 antagonist) have been discovered in various modifications both for GPCR agonists as well as antagonists. Figure 2 exemplifies six different spirpoperidine scaffolds identified for six different G-protein coupled receptors.

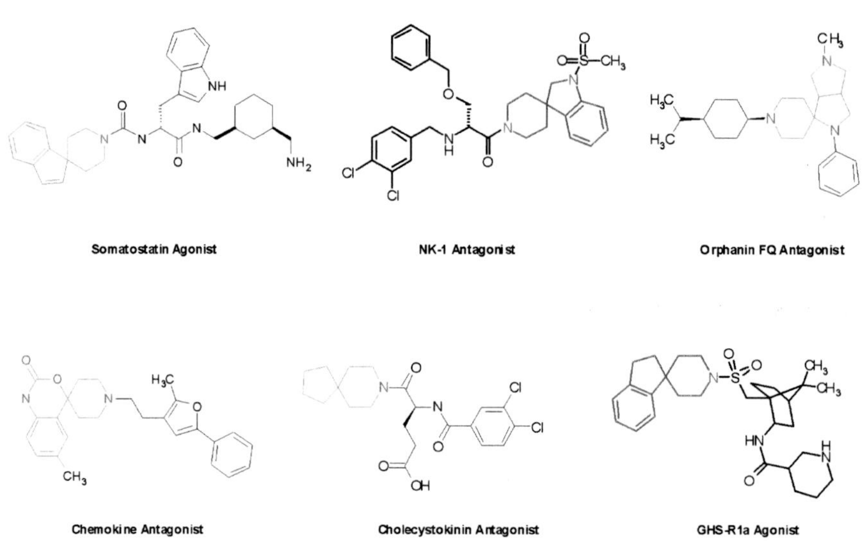

Figure 2. Reported GPCR ligands containing spiropiperdines as recognition motives.

While 'privileged structures' are supposed to show some promiscuity within a protein family, the terminus 'needle' was described as a molecular fragment of ligands binding specifically to particular protein family members. One example for such a needle within the GPCR area is the ortho-substituted biphenyl tetrazole. This motif is well known as a fragment of AT-1 ligands and can therefore be regarded as an 'AT-1 needle' since it appears in most angiotensin-1 antagonists reported in the literature [8].

For the identification of novel small molecule ligands targeting the neurokinin-1 receptor we initiated the generation of focused libraries based on the design strategy to combine the two concepts of 'privileged structures' and 'needles' [9,10]. All three known neurokinin receptors (NK-1, NK-2 & NK-3) belong to the target family of 7-transmembrane G-protein coupled receptors (family 1b). The NK-2 receptor is mainly expressed in the periphery whereas the NK-1 and NK-3 receptors are mainly expressed in the central nervous system indicating their potential therapeutic utility ranging from CNS indications to respiratory and gastric diseases.

The endogenous ligands for the neurokinin receptors are the tachykinins, a group of peptides that share a common C-terminal amino acid sequence Phe-X-Gly-Leu-Met-NH_2 where X is either Phe or Val. The most prominent member of this peptide family is the undecapeptide 'Substance P' (X = Phe) which shows highest affinity for the NK-1 receptor, whereas NKA and NKB (X = Val) are both decapeptides that bind preferentially to the NK-2 and NK-3 receptor, respectively.

Although GPCRs with such large peptide ligands as natural substrates are supposed to be rather difficult to be modulated by small molecules, several drug like NK-1 receptor modulators have been reported in the literature, some representatives of which are depicted in Fig. 3.

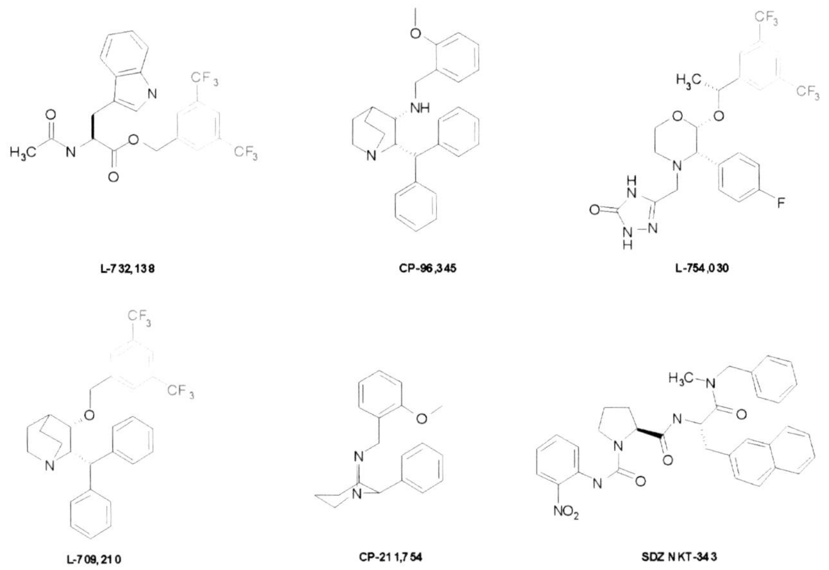

Figure 3. Reported small molecule NK-1 receptor ligands.

As indicated in Fig. 3, many NK-1 ligands have been reported that do possess the 3,5-bis-(trifluoromethyl)phenyl group (highlighted in red) which is meanwhile well recognized as an 'NK-1 needle'. The combination of promiscuous 'privileged structures' and the fairly target specific 'NK-1 needle' gave rise to a targeted library design where the remaining exit vectors in the scaffold (spiropiperidines) were decorated randomly (Figs 4 and 5).

Instead of extensively exploiting the chemotypes by analogizing each exit vector in the templates simultaneously we decided to either concentrate on one additional modification or employ only relatively small building blocks to keep the molecular weight and the resulting lipophilicity in an appropriate range.

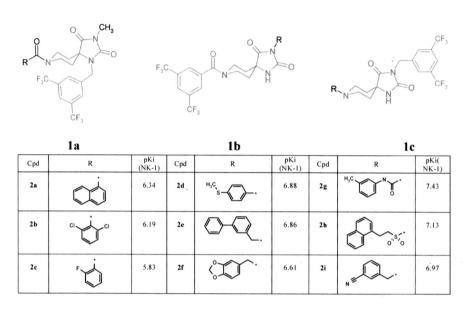

Cpd	R	pKi (NK-1)	Cpd	R	pKi (NK-1)	Cpd	R	pKi(NK-1)
2a		6.34	2d		6.88	2g		7.43
2b		6.19	2e		6.86	2h		7.13
2c		5.83	2f		6.61	2i		6.97

Figure 4. Spirohydantoins as novel NK-1 receptor ligands.

The first compound arrays were based on the spirohydantoin template. These compounds can be generated rapidly from the corresponding amino acid (4-Amino-piperidine-4-carboxylic acid). The 'NK-1 needle' was either introduced *via* the 3,5-bis(trifluoromethyl)benzyl chloride at the α-amino acid nitrogen **1a**, 3,5-bis(trifluoromethyl)benzoic acid at the piperidine nitrogen **1b**, or the 3,5-bis(trifluoromethyl)phenyl- or 3,5-bis(trifluoromethyl)benzyl isocyanate at the imide nitrogen **1c**, respectively.

For all three sublibraries, ligands were identified that showed decent affinities ($pK_i > 5$) in a radio-ligand displacement assay. The three most active ligands for each sublibrary are disclosed in Fig. 4 (**2a-i**). From a set of 136 compounds, 97 molecules showed a binding affinity of $pK_i(hNK-1) > 5$ which corresponds to a hit rate of 71%.

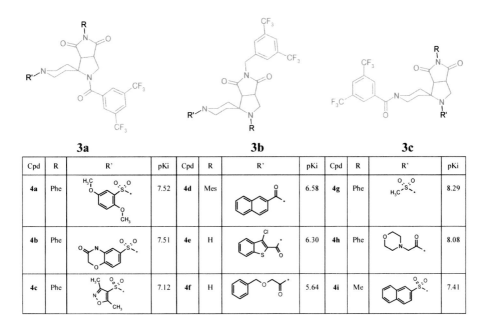

Cpd	R	R'	pKi	Cpd	R	R'	pKi	Cpd	R	R'	pKi
4a	Phe		7.52	4d	Mes		6.58	4g	Phe		8.29
4b	Phe		7.51	4e	H		6.30	4h	Phe		8.08
4c	Phe		7.12	4f	H		5.64	4i	Me		7.41

Figure 5. Spiropyrrolo-pyrrols as novel NK-1 receptor ligands.

As a second example we started from the correspondingly orthogonal protected or resin-bound spiropyrrolo-pyrrole. This template was identified in the orphanin FQ (OFQ) area, a target protein that also belongs to the super-family of peptide class 1 GPCRs. Again the fairly specific 'NK-1 needle' was introduced into the scaffold at all three possible positions (**3a-c**) and the remaining vectors decorated using solution- and/or solid-phase chemistry.

As already observed for the spirohydantoins, independently from the position of the 'NK-1 needle' within the spiropyrrolo-pyrrole template, various low nanomolar binding molecules were identified. From an array of 132 compounds submitted for testing, 91 ligands were identified with a binding affinity of $pK_i(hNK-1) > 5$, resulting in a hit rate of 69%. Nine representative hits are depicted in Fig. 5 (**4a-i**).

The two examples discussed above exemplify the application of 'privileged structure' based library design for the generation of small molecule NK-1 ligands. As powerful as this approach can be it is obvious that the search for novel 'privileged structures' and/or 'needles' is essential. *De novo* design tools such as Skelgen [11] or TOPAS [12] have been described that allow so-called 'scaffold hopping', a method of moving into novel and therefore patent free chemotypes.

TARGETED LIBRARIES FROM VIRTUAL SCREENING

In contrast to the physical high-throughput screening approach compound filtering and clustering on a computational basis allows the elimination of undesired structures in terms of chemically reactive functionalities, predicted liabilities (e.g. frequent hitter, hERG, CYP450), drug-like properties etc. or group them based on certain similarity/diversity criteria [13]. This is most often applied in the context of compound purchasing to ensure the quality of molecules to be brought in and also to prevent duplication of topologically similar structures already represented in the compound inventory.

A further step downstream the drug discovery chain is the application of such virtual screening tools to pre-select compounds from the corporate collection for directed or biased screening efforts. This is usually applied when a high-throughput screening assay is not available or not appropriate. Several retrospective screening analyses have shown the validity of such an approach where compounds were predicted to be active and increased hit rates were finally observed. To a lesser extent prospective experiments are described showing that biased screening efforts can lead to novel hits where an HTS campaign did not deliver reasonable structures. The integration of virtual screening and HTS seems to be a logical consequence of the maturation of both disciplines but is often still regarded as being competitive rather then complementary and therefore not as intensively applied in many drug discovery companies as one might expect [14].

For the *de novo* generation of targeted libraries basically the same type of algorithms can be applied by computationally screening not physically available compound collections, but rather virtual libraries. Since the number of available structures from such libraries is essentially infinite, the tools and strategies to be applied for computational screening often differ from those used for physically available compound collections. Usually a whole cascade of virtual screening algorithms ranging from 1D to 3D tools are applied to narrow down the huge number of theoretically available compounds to some 100-1000 predicted actives [15].

We applied TOPAS (a 2D topological screening algorithm) as a *de novo design* tool in our search for novel cannabinoid receptor ligands [16]. These receptors also belong to the class 1 GPCR family where two subtypes are currently known. The centrally expressed protein (CB-1) is supposed to modulate appetite by the binding of Anandamide, a lipid GPCR ligand, making it a promising novel drug target for obesity and related diseases.

Starting from a pool of ~ 1500 GPCR fragments (generated from a set of known GPCR ligands) *de novo* designs were generated by linking such fragments randomly and comparing the resulting virtual molecules with known CB-1 ligands based on their topological similarity (CATS) [17]. Two of the most promising proposals are depicted in Fig. 6. Besides the drug-like 'look' of such molecules their chemical tractability is of utmost importance to allow the very rapid validation of the proposed structures. A parallel synthesis approach must be feasible to compensate the fuzziness of the design proposals by generating and testing compound arrays rather than single molecules.

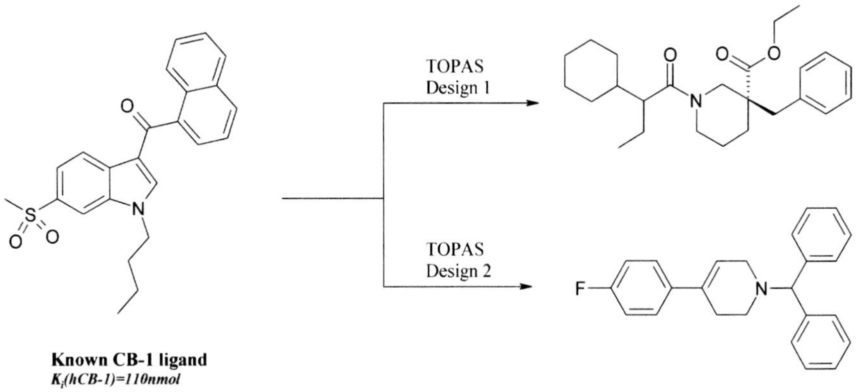

TOPAS
Design 1

TOPAS
Design 2

Known CB-1 ligand
K_i(hCB-1)=110nmol

Figure 6. 'Scaffold hopping' proposals generated de novo by TOPAS.

For Design 1 a small array of 83 analogues was prepared and tested in a radio-ligand binding assay. Five hits were identified corresponding to a hit rate of 6%. Fifty analogues of Design 2 were generated, again delivering five hits corresponding to 10% hit rate. Based on the chemical tractability for both design proposals follow-up libraries were generated rapidly to give compound series with low nanomolar binding affinity, the desired functionality (inverse agonists) and preferable physico-chemical properties. In addition, the compounds were further investigated concerning their microsomal stability and cytochrome P_{450} interactions, indicating no issues in this regard. To exclude polypharmacology, a representative lead structure was further tested against CB-2 binding as well as 80 unrelated protein targets before moving into the lead optimization phase.

This 'scaffold hopping' concept can significantly speed up the hit and lead generation process by efficiently combining the advantages of virtual screening and rapid combinatorial chemistry.

The whole project consisted of four consecutive steps: i) fragment-based *de novo* design of virtual analogues by TOPAS; ii) selection of preferred building blocks for synthesis; iii) parallel synthesis, purification and characterization of two small compound libraries and subsequently, iv) biological testing for activity in a human CB-1 receptor binding assay. Three representatives of each chemotype are depicted in Fig. 7.

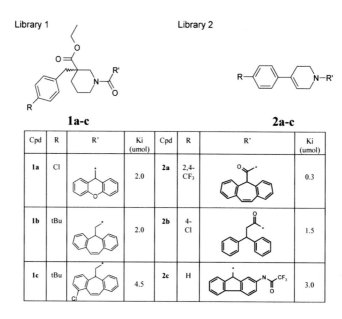

Figure 7. Representative CB-1 ligands derived from design proposals 1 & 2 (Figure 6).

Essentially the same strategy was applied to a virtual library where molecular structures are not generated on a randomly combined fragment-basis but rather on templates and building blocks defined by the chemist (RADDAR: Roche Adaptive Drug Design And Refinement). Such libraries were computationally screened to generate predicted hits where, in contrast to the *de novo* approach, the chemistry is already established and the building blocks readily available. Again small arrays were generated based on a the same topological similarity algorithm (CATS) using known CB-1 ligands as seed structures to result in compound libraries that showed much higher hit rates (> 10%) compared to random screening approaches (structures not shown).

Figure 8. RADDAR: a virtually infinite number of chemical structures can be generated by computationally enumerating combinatorial library proposals. 'Cherry picking' of interesting candidates (proposed structures) based on defined computational algorithms allows the chemist to synthesize these subsets of theoretically accessible molecules that are predicted to be of relevance.

CONCLUSION

Since the generation of lead compounds encompasses much more than affinity for particular targets, it is essential that all the early phase drug discovery issues are addressed. Multidimensional Optimization (MDO), the simultaneous investigation of affinity, selectivity, function, molecular properties and DMPK (*in vitro* and *in vivo*) has been set up in many pharmaceutical research organizations to ensure that compound series with potential liabilities are identified early so as to concentrate on candidates with an appropriate profile, instead of only optimizing affinity in the hope of subsequently finding a remedy for other issues (such as solubility or drug-drug interactions etc.).

It is here that novel methodologies like computational prediction algorithms can have a major impact since the generation of a physico-chemical properties and ADMET profile for whole compound libraries is not only time consuming but also very costly. Although the level of predictive accuracy for ADMET profiles is still suffering from the lack of datasets associated with chemical structures, huge efforts are underway to gain more confidence in those prediction tools which will allow a more informed decision making process when lead series have to be evaluated for further optimization studies and clinical candidate selection [18].

REFERENCES

[1]   Drews, J. (2000) Drug discovery: a historical perspective. *Science* **287**:1960-1963.

[2]   Watson, S. Arkinstall, S. (1994) *The G-Protein Linked Receptors Facts Book.* Academic Press, New York.

[3]   Flower, D.R. (1999) Modeling G-protein-coupled receptors for drug design. *Biochem. Biophys. Acta* **1422**:207-234.

[4]   Bleicher, K., Green, L., Martin, R., Rogers-Evans, M. (2004) Ligand identification for G-protein coupled receptors: a lead generation perspective. *Curr. Opin. Chem. Biol.* **8**:287-296.

[5]   Bondensgaard, K., Ankernsen, M., Thorgensen, H., Hansen, B., Wulff, B., Bywater, R. (2004) Recognition of privileged structures by G-protein coupled receptors. *J. Med Chem.* **47**:888-899.

[6]   Perlman, J., Colson, A.-O., Jain, R., Czyzewski, B., Cohen, C., Osman, R., Gershengorn, M. (1997) Role of the extracellular loops of the thyrotropin-releasing hormone receptor: evidence for an initial interaction with thyrotropin-releasing hormone. *Biochemistry* **36**:15670-15676.

[7]   Evans, B.E. *et al.* (1988) Methods for drug discovery: development of potent, selective, orally effective cholecystokinin antagonists. *J. Med. Chem.* **31**:2235-2246.

[8]   Burnier, M., Brunner, H. (2000) Angiotensin II receptor antagonists. *Lancet* **355**:637-645.

[9]   Bleicher, K., Wüthrich, Y., Deboni, M., Kolczewski, S., Hoffmann, T., Sleight, A. (2002) Parallel solution- and solid-phase synthesis of spirohydantoin derivatives as neurokinin-1 receptor ligands. *Bioorg. Med. Chem. Lett.* **12**:2519-2522.

[10]   Bleicher, K., Wüthrich,Y., Adam, G., Hoffmann, T., Sleight, A. (2002) Parallel solution- and solid-phase synthesis of spiropyrrolo-pyrroles as novel neurokinin receptor ligands. *Bioorg. Med. Chem. Lett.* **12**:3073-3076.

[11]   Lloyd, D., Bünemann, C., Todorov, N., Manalack, D., Dean, P. (2004) Scaffold hopping in *de novo* design. Ligand generation in the absence of receptor information. *J. Med.Chem.* **47**:493-496.

[12]   Schneider, G., Clément-Chomienne, O., Hilfiger, O., Schneider, P., Kirsch, S., Böhm, H.-J. Neidhart, W. (2000) Virtual screening for bioactive molecules by evolutionary de novo design. *Angew. Chem. Int. Ed. Engl.* **39**:4130-4133.

[13]   Böhm, H.J., Schneider, G. (Eds) (2000) *Virtual Screening for Bioactive Molecules.* Wiley- VCH, Weinheim.

[14]   Bajorath, J. (2002) Integration of virtual and high-throughput screening. *Nature Rev. Drug Discov.* **1**:882-894.

[15]   Walters, W. P., Stahl, M. T., Murcko, M. A. (1998) Virtual screening - an overview. *Drug Discov. Today* **3**:160-178.

[16]   Rogers-Evans, M., Alanine, A. I., Bleicher, K.H., Kube, D., Schneider, G. (2004) Identification novel cannabinoid receptor ligands via evolutionary de novo design and rapid parallel synthesis. *QSAR Comb. Sci.* **23**:426-430.

[17]   Schneider, G., Neidhart, W., Giller, T., Schmid, G. (1999) "Scaffold-hopping" by topological pharmacaphore search: a contribution to virtual screening. *Angew. Chem. Int. Ed. Engl.* **38**:2894-2896.

[18]   Davis, A., Riley, R.J. (2004) Predictive ADMET studies, the challenges and the opportunities. *Curr. Opin. Chem. Biol.* **8**:386-386.

 Beilstein-Institut

The Chemical Theatre of Biological Systems, May 24th - 28th, 2004, Bozen, Italy

SELECTIVE CHEMICAL INTERVENTION IN BIOLOGICAL SYSTEMS: THE SMALL MOLECULE TOOL, (*S*)-(-)-BLEBBISTATIN

STEPHEN PATTERSON, CHRISTINA LUCAS-LOPEZ
AND NICHOLAS J. WESTWOOD

School of Chemistry and Centre for Biomolecular Sciences, University of St Andrews, North Haugh, St Andrews, Fife, Scotland, KY16 9ST, U.K.
E-Mail: *njw3@st-andrews.ac.uk

Received: 17th January 2005 / Published: 22nd July 2005

ABSTRACT

Selective small molecule inhibitors of protein function provide a method of studying biological processes that is often complementary to classical genetic and RNAi-based approaches. This article focuses on a recently identified small molecule known as blebbistatin. We review blebbistatin's discovery, biological characterization, selectivity and continuing use. Synthetic chemistry has played a key role in the blebbistatin story and we also review our recent work relating to the asymmetric synthesis and absolute stereochemical assignment of the active enantiomer, (*S*)-(-)-blebbistatin. High-throughput synthetic approaches to blebbistatin analogues are discussed and a novel analogue is described that has significantly improved physical properties for use in fluorescence-based imaging experiments on live cells. This article looks to emphasize the multidisciplinary nature of research projects in chemical genetics.

INTRODUCTION

The search for novel small molecule modulators of protein function continues to gather momentum in academia. For example, the NIH roadmaps in "accelerating medical discovery", "building blocks, biological pathways and networks" and "molecular libraries and imaging" provide a medium- to long-term view of the US commitment to research in chemical genetics (defined as a discovery platform using small molecules that alter the function of specific proteins in place of mutations).

This revival in the use of small molecules in academic biology has occurred because they provide a method of dissecting biological processes that is often complementary to classical genetics and RNAi technology. Interestingly, it seems that there is even a place for small molecules in the emerging field of "systems biology". The World Technology Evaluation Centre (WTEC) in its list of the technologies required for systems biology research includes *'tools for analyzing the spatial and temporal behaviour of networks'*. Two recent review articles support this view [1,2]. The ability to reversibly control the activity of a protein as a function of time and location is a key advantage of small molecules. This is achieved by "washing in" or "out" the small molecule at a chosen time in a chosen cell type - both factors that are under the researcher's control. "Caging" the small molecule can also provide the required level of control [3]. Figure 1 lists several other reasons why novel small molecules are of importance in post-genomic science.

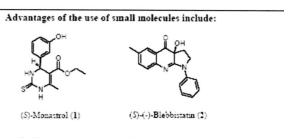

Advantages of the use of small molecules include:

(S)-Monastrol (1)   (S)-(-)-Blebbistatin (2)

1. Easily applicable to cross-species studies.
2. Used to study systems where standard genetic tools are unavailable, rudimentary or difficult to use.
3. Study of the function of essential or recessive genes.
4. Excellent temporal resolution (particularly important in the study of rapid cellular processes).
5. The ability to observe biological responses following release from a small molecule block (due to reversibility of action of most small molecules).
6. Use of small molecules in combination (multiple knock-outs/ins).
7. Can be used in most experimental situations including *in vitro* (e.g. purified protein studies), in the cell type of the researcher's chosing or *in vivo*.
8. Precise temporal and spatial control of activity using caging techniques.
9. Potentially close link to drug discovery.

Figure 1. Small molecules and chemical genetics: *Forward chemical genetics* involves identifying a phenotype in an organism or cell caused by a small molecule and then identifying the protein target (c.f. forward genetics): *Reverse chemical genetics* involves selection of a protein target of interest and screening for small molecules that affect the protein's activity (c.f. reverse genetics). All definitions relating to chemical genetics are taken from http://iccb.med.harvard.edu.

The recently identified small molecule tool monastrol (**1**) provides an excellent example of the impact that a selective modulator of protein function can have [4,5].

In this article, we review the literature surrounding another example that is having an equally dramatic influence on both the biology and chemistry communities. Since its first report in March 2003 [6], 22 peer-reviewed research papers (as of 4th April 2005) have been published focusing on the use of the small molecule tool, (*S*)-(-)-blebbistatin (**2**).

IDENTIFICATION OF (±)-BLEBBISTATIN BY HIGH-THROUGHPUT SCREENING

Small molecules that modulate the activity of cytoskeletal proteins are of great use to cell biologists [7]. For example latrunculin A, a marine natural product that binds monomeric actin [8] and prevents its incorporation into filaments, has been used extensively to study the cellular role of actin (for a review see [7]). However, there are many areas of cytoskeletal research that lack the necessary small molecule tools. For example, the detailed study of cytokinesis in mammalian cells would benefit from a small molecule that is capable of preventing ingression of the cleavage furrow *without* preventing furrow assembly. Halting the cell cycle in cytokinesis can be achieved by inhibiting the action of *non-muscle* myosin II (NMII) [9]. Unfortunately the widely used (and controversial [10]) *muscle* myosin II inhibitor, BDM [11], has been reported not to inhibit NMII [12]. Therefore, in order to study cytokinesis in more detail using a chemical genetic approach it was necessary to *discover* a novel small molecule that inhibits the ATPase activity of NMII.

Mitchison and co-workers employed high-throughput screening (HTS) of commercially available chemical libraries to identify inhibitors of NMIIA ATPase activity [6], a strategy that had already been successfully used in their laboratory to discover novel inhibitors of muscle myosin II [12]. From over 16,000 screening compounds, 4 inhibitors of NMIIA ATPase activity were identified (Fig. 2, panel B), of which (±)-blebbistatin was the most potent.

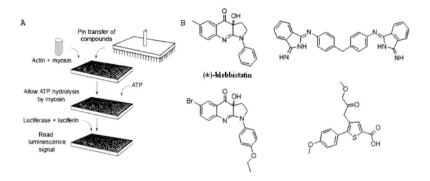

Figure 2. A reverse chemical genetic approach to discover novel small molecule tools: **A)** Diagrammatic representation of the coupled assay procedure used to identify inhibitors of non-muscle myosin IIA (one of three isoforms that exist). The assay was performed by adding DMSO solutions of small molecules to wells containing a solution of actin and human platelet NMIIA. The assay was performed in a 1536 well format, with nL volumes of solutions being transferred using an automated pin transfer device. The assay was then initiated by the addition of ATP. The assay plates were incubated to allow ATP hydrolysis to occur prior to the addition of a solution of luciferase and luciferin. The resulting luminescence was then recorded as a direct measure of the inhibitory effect of each small molecule; the greater the luminescence the more effective an inhibitor the small molecule was. **B)** The four inhibitors of NMIIA identified by HTS including (±)-blebbistatin [6]. Copyright Nature Publishing Group, http://cellbio.nature.com [12].

PRELIMINARY BIOLOGICAL CHARACTERIZATION OF (±)-BLEBBISTATIN

As with all reverse chemical genetic approaches, the next stage of this programme was to determine the effect of (±)-blebbistatin upon NMII-dependent processes in cells. (±)-Blebbistatin at a concentration of 100 mM was found to inhibit cell **blebb**ing in M2 cells (explaining the compound's name) and to perturb cell motility in vertebrate cells (goldfish keratocyte assay) [6]. Consistent with the goals of the study, 100 mM (±)-blebbistatin was also found to block cleavage furrow contraction in dividing *Xenopus* tissue culture (XTC) cells [6]. Importantly correct assembly of the cleavage furrow occurred in the presence of (±)-blebbistatin, with the localization of non-muscle myosin II, anillin (another component of the cleavage furrow [13]) and microtubules being unaffected. All of the inhibitory effects of (±)-blebbistatin were found to be reversible, the cells re-entering the cell cycle after 'washing out' of the inhibitor.

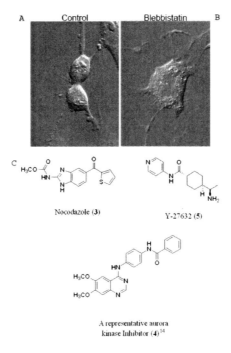

Nocodazole (3)

Y-27632 (5)

A representative aurora
kinase Inhibitor (4)[14]

Figure 3. The effect of (±)-blebbistatin on cytokinesis. **Image A** shows an XTC cell undergoing normal cell division in the absence of drug treatment (control cell), note the presence of the contracted cleavage furrow. **Image B** shows the same cell type treated with 100 mM (±)-blebbistatin, Note the complete absence of cleavage furrow contraction despite successful chromosome segregation. Movies showing this process in more detail are also available in the supplementary material for [6]. (http://www.sciencemag.org/cgi/content/full/299/5613/1743/DC1). **C)** The structures of nocodazole (3), aurora kinase inhibitor (4) and Y-27634 (5). These small molecules were used in conjunction with (±)-blebbistatin to carry out double chemical knockout experiments. Copyright Science Magazine.

The ability of (±)-blebbistatin to block furrow cleavage without affecting the localization of other cytoskeletal components allowed 'double chemical knockout experiments' to be carried out. This involved incubating cells in the presence of both (±)-blebbistatin and other small molecule tools. For example, incubation of (±)-blebbistatin-arrested HeLa cells with nocodazole (3) (a small molecule tubulin depolymerizer) resulted in cells with delocalized non-muscle myosin II (NMII) and anillin and with no microtubule mid-zone, thus demonstrating an essential role for microtubules in maintaining furrow component localization. An analogous experiment in which nocodazole (3) was replaced by an aurora kinase inhibitor (e.g. 4) resulted in the delocalization of NMII *but not anillin*. The mid-zone microtubules were also disrupted. This result provided the first evidence that the localization of these two furrow components is *independently regulated*.

A further experiment involving treatment of (±)-blebbistatin-arrested cells with the rho-kinase inhibitor, Y-27634 (**5**), also led to delocalization of NMII and not anillin, but this time had no effect upon the mid-zone microtubules.

It appears therefore that the action of rho-kinase is required for localization of NMII into the cleavage furrow and that aurora kinase is required for mid-zone microtubule organization. Following this initial report on the use of (±)-blebbistatin, several other groups have used this small molecule tool to study other NMII-dependent cellular process. This literature is briefly reviewed in the section entitled (*S*)-(-)-Blebbistatin (**2**) as a Molecular Tool.

THE CHEMICAL STRUCTURE OF (±)-BLEBBISTATIN

A key step in HTS approaches using commercially available small molecule collections is to confirm the structure of the initially identified hit. In the case of (±)-blebbistatin, this process proved particularly interesting as (±)-blebbistatin itself was not present in the commercial collection that was used. In fact, the compound present in the "(±)-blebbistatin hit well" was listed as **6** (Fig. 4). A repurchased and freshly dissolved sample of **6** did not inhibit NMII.

The relatively rapid conversion of a DMSO stock of **6** from a colourless to bright yellow solution suggested, however, that **6** degrades to produce the NMII inhibitor. Whilst not the first time that a degradation product or contaminant of the intended inhibitor has proved to be the bioactive component [15], these cases are relatively rare. Initial studies carried out in the Mitchison laboratory, showed that an "aged" DMSO solution of **6** does indeed degrade to (±)-blebbistatin (Fig. 4).

The 'degradation' of **6** to (±)-blebbistatin could occur via quinolone **7** followed by reaction with singlet oxygen, analogous to our synthetic route [16]. Evidence in support of this hypothesis could be obtained by conducting experiments with $[^{18}O_2]$-oxygen gas and these studies are ongoing in our laboratory. Subsequent studies from the Mitchison and Westwood laboratories have provided a more efficient route to (±)-blebbistatin via quinolone **7** [16].

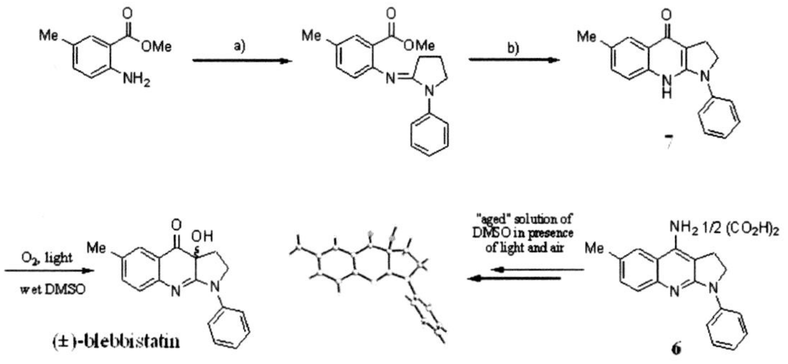

Figure 4. Preparation of (±)-blebbistatin. Reagents and conditions: a) i) $POCl_3$, CH_2Cl_2, 25°C, 3 h; ii) *N*-phenyl-2-pyrrolidinone, 40°C, 16 h, 41%; b) LiHMDS (3 equiv), -78°C to 0°C, 3 h, 90%; c) O_2, hv, rose Bengal, DMSO, 25°C, 3 h, 29%; or O_2, hv, rose Bengal, 25°C, 3 h, THF, 26%. LiHMDS = lithium bis(trimethylsilyl)amide [16]. X-ray crystallographic analysis of a sample of (±)-blebbistatin confirmed the atom connectivity is as shown.

(±)-Blebbistatin possesses a single chiral centre. It is well established that one enantiomer of a bioactive compound is often more potent than the other, or even that one enantiomer is inactive or induces a different unrelated phenotype. Therefore, Mitchison and co-workers separated the enantiomers of (±)-blebbistatin using chiral HPLC techniques and subjected them to individual biological testing [6]. *In vitro* assays with (-)-blebbistatin (**2**) (the enantiomer that rotates the plane of plane polarized light in the -ve direction) gave an IC50 value of ~2 mM against NMIIA, whilst (+)-blebbistatin (**8**) (Fig. 5) was inactive. Cellular assays performed with (-)-2 or (+)-blebbistatin (**8**) demonstrated that only (-)-blebbistatin (**2**) was able to arrest cells in cytokinesis, consistent with the trend established *in vitro*. Whilst a report [17] has called into question these results, it is now clear that the discrepancies arise from the lower levels of optical purity associated with some commercially available sources of **2** and **8**. In the light of these results, it became desirable to develop a highly efficient route to optically pure samples of both **2** and **8**.

The selective formation of chiral centres presents a significant challenge to the organic chemist. However, precedent exists for the use of *N*-sulfonyloxaziridines such as **9** and **10** (Fig. 5) to perform asymmetric hydroxylations of ketone enolates in high yield and with excellent enantiomeric excesses (e.e., the percentage difference in the abundance of 2 enantiomers formed in a reaction e.g. if the enantiomer ratio was 95:5 the e.e. would be 90%) [18,19]. It was envisaged that optically enriched (-)-2 or (+)-blebbistatin (8) could be prepared from quinolone 7 using this chemical methodology (Fig. 5) [16].

This late stage oxidation was attractive as both enantiomers could be prepared in a single step from a common intermediate. Treatment of the lithium enolate of quinolone **7** with oxaziridine **9** gave optically enriched (-)-blebbistatin (**2**) in moderate e.e. (entry 2, Fig. 5). The reaction temperature, oxaziridine, and base were all varied in order to identify optimal conditions for the formation of (-)-blebbistatin (**2**). These studies resulted in a procedure that enables the synthesis of highly optically enriched (>99.5% e.e.) **2** after a single recrystallization step (entry 4).

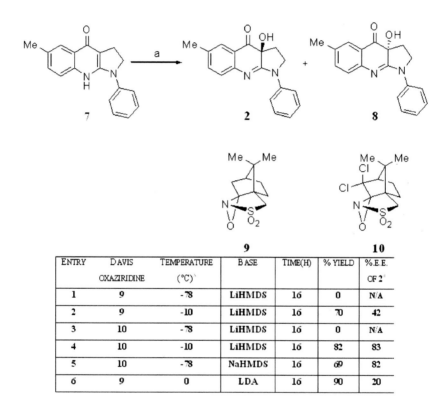

ENTRY	DAVIS OXAZIRIDINE	TEMPERATURE (°C)	BASE	TIME(H)	% YIELD	%.E.E. OF 2
1	9	-78	LiHMDS	16	0	N/A
2	9	-10	LiHMDS	16	70	42
3	10	-78	LiHMDS	16	0	N/A
4	10	-10	LiHMDS	16	82	83
5	10	-78	NaHMDS	16	69	82
6	9	0	LDA	16	90	20

Figure 5. Optimization of asymmetric hydroxylation of **7** to give enantiomerically enriched (-)-(**2**) using the Davis oxaziridine methodology [16]. The hydroxylation reaction was performed using different bases; LiHMDS (lithium bis-(trimethylsilyl)amide), NaHMDS (sodium bis-(trimethylsilyl)amide) and LDA (lithium di-isopropylamide). Various reaction temperatures and two commercially available oxaziridines were also tested in order to determine the optimal reaction conditions. The e.e. values were determined using chiral HPLC analysis of the crude reaction mixture.

Neurological and cancer biology studies using (±)-blebbistatin have identified myosin II as a putative drug target (see the section entitled (*S*)-(-)-Blebbistatin (**2**) as a Molecular Tool) [20,21]. As a result (-)-blebbistatin (**2**) could be considered as a lead compound for pharmaceutical development.

An efficient route to highly optically enriched (-)-blebbistatin (**2**) and its analogues is therefore of additional significance given the high level of optical purity typically required in even potential therapeutics. Preliminary studies from our group indicate that whilst subtle variations in e.e. do occur, this methodology is very tolerant of substitutions in either aromatic ring in **7**.

In addition to being able to synthesize small molecule tools in the required purity levels, it is also necessary to determine their absolute stereochemistry (where relevant). Determining if the bioactive enantiomer is *R* or *S* aids computer modelling-based studies designed to identify (-)-blebbistatin's (**2**) myosin binding site [17]. Knowing the absolute stereochemistry is also essential in the planning of alternative synthetic routes to (-)-blebbistatin (**2**) and its analogues. As (-)-blebbistatin (**2**) contains no "heavy atoms", it is difficult to determine its absolute stereochemistry using X-ray diffraction techniques in the absence of extremely high quality crystals. In order to overcome this problem, a heavy atom (bromine)-containing analogue **11** was prepared (Fig. 6). X-ray crystallographic analysis of **11** showed that its absolute stereochemistry was *S*. Subsequent replacement of the bromine in (*S*)-(-)-11 with a hydrogen atom using catalytic hydrogenation gave exclusively (-)-blebbistatin (**2**) (confirmed by comparison with authentic material using chiral HPLC). Therefore, *the absolute stereochemistry of (-)-blebbistatin (**2**) is S*. Further evidence in support of this conclusion comes from a recently published X-ray crystal structure of (*S*)-(-)-blebbistatin (**2**) bound to the motor domain of *Dictyostelium discoideum* myosin II (see section entitled Future Studies).

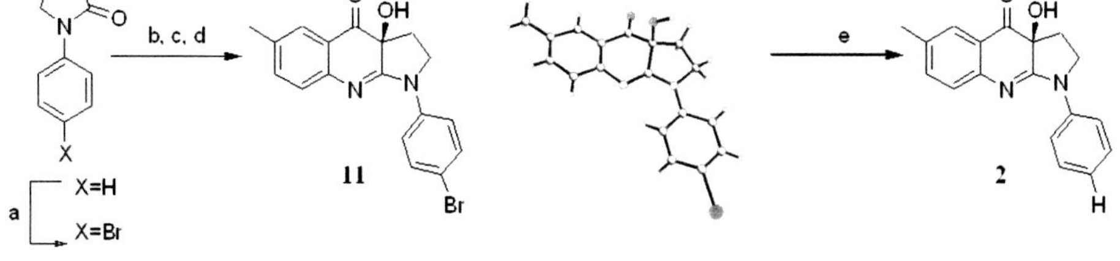

Figure 6. Assignment of the absolute stereochemistry of (*S*)-(-)-blebbistatin (**2**). Reagents and conditions: a) NBS, DMF, 25°C, 2 days, 50%; b) i) POCl$_3$, CH$_2$Cl$_2$, 25°C, 3 h; ii) methyl 5-methylanthranilate, 40°C, 16 h, 26%; c) LiHMDS (3 equiv), -78°C to 0°C, 3 h, 60%; d) i) LiHMDS (1.2 equiv), THF, -78°C; ii) **10** (2.4 equiv), -10°C, 16 h, 68%, 88% *ee*; recrystallization from acetonitrile >99% e.e., [α]D$_{26}$ -526 (c=0.1, CH$_2$Cl$_2$); f) NBS = *N*-bromosuccinimide, DMF = *N,N*-dimethylformamide.

ASSESSING THE SELECTIVITY OF SMALL MOLECULE TOOLS

The most frequent criticism of the use of small molecules to study biological processes is that they lack sufficient selectivity to enable "clean" biological questions to be asked. With the increase in methods [22] of assessing selectivity, the onus is increasingly on the developer of novel small molecule tools to demonstrate just how selective the latest addition to the toolbox is. In this case, a large superfamily of myosins that function as actin-dependent motor proteins is known to exist. Each family member possesses a globular motor domain [23]. A recent phylogenetic analysis using the motor domain sequences of known or predicted myosins from several species divided the superfamily into 18 classes [24]. Additionally, some myosin classes are further divided into subclasses [25]. For example the myosin II class is partitioned into subclasses that include skeletal, cardiac, smooth muscle and non-muscle. Each subclass may contain numerous isoforms (for example nonmuscle myosin II exists as three isoforms A, B and C) [26]. Assessing selectivity is a daunting (but essential) task even when it is assumed that no other protein families are being "unintentionally targeted". The state of the art is demonstrated by studies on kinase inhibitors [27].

Sellers and co-workers have conducted a programme of *in vitro* ATPase assays to determine which myosins are inhibited by (±)-blebbistatin [28]. The ability of (±)-blebbistatin to inhibit proteins from the myosin classes I, V, X and XV was determined. None of these 'unconventional' myosins were inhibited to any significant degree, even at (±)-blebbistatin concentrations of 100 mM [28]. Although only 5 myosin classes were assayed, the currently available data suggest that (±)-blebbistatin and hence (S)-(-)-blebbistatin (2) appears to be a *specific* inhibitor of myosin class II.

The activity of (±)-blebbistatin against different myosin II subclasses has also been investigated [28]. Rabbit *skeletal* muscle myosin, porcine β-*cardiac* muscle myosin, human NMIIA and chicken NMIIB are all inhibited by (±)-blebbistatin to a similar extent (IC_{50} values of 0.5, 1.2, 5.1 and 1.8 mM respectively, Fig. 7), demonstrating limited selectivity within the myosin II class and between isoforms within the same subclass (NMII). However, turkey smooth muscle myosin (closely related to NMII) is inhibited by (±)-blebbistatin with an IC50 value of 79.6 mM, significantly higher than that for NMIIA and NMIIB [28]. This demonstrates that (±)-blebbistatin does possess a degree of selective inhibition even within subclasses. However, there is considerable room for improvement providing an exciting (and challenging) opportunity for synthetic chemists who can identify subclass or isoform selective analogues.

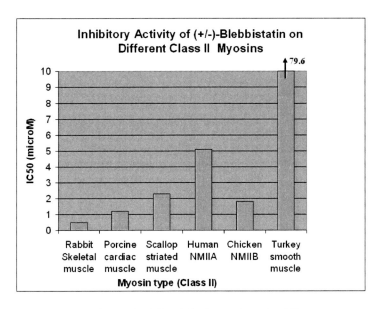

Figure 7. Assessment of (±)-blebbistatin selectivity within the myosin II class [28].

In a recent study, Ivanov and co-workers [29] were only able to determine that NMIIA was involved in the disassembly of the apical junction complex (AJC) (see following section) by determining the expression levels of the three NMII isoforms using specific antibodies. This extended set of experiments would not have been necessary if isoform selective inhibitors were available. Additionally, if the cell type used in these studies expressed two or more isoforms of NMII at similar levels no conclusion could have been drawn as to which isoform was responsible for the observed phenotype.

In addition to the myosin II class (see previous section), the unconventional myosin classes also represent important drug targets. For example class XIV consists of myosins from the *Apicomplexan* parasites *Toxoplasma gondii* and *Plasmodium falciparum* [24], both of which are important human pathogens. TgMyoA, a member of this class, has been shown to play an essential role in parasitic motility *in vitro* and in cell invasion *in vitro* and *in vivo* [30]. It can therefore be considered as a validated drug target and selective myosin XIV inhibitors as potentially novel therapeutics. In addition, specific small molecule inhibitors of host cell invasion by *T. gondii* are of use to biologists, due to the fact that the obligate intracellular nature of these parasites renders TgMyoA knockout lethal.

A recently reported HTS [31] identified a number of small molecules that inhibit host cell invasion by *T. gondii*, although it remains unclear whether any of these small molecules target TgMyoA.

(*S*)-(-)-BLEBBISTATIN (2) AS A MOLECULAR TOOL

To meet the needs of the cellular biology community (±)-, (*S*)-(-)-(2) and (R)-(+)-blebbistatin (**8**) are commercially available. Their availability has enabled researchers to investigate the role of subclasses of myosin II in cellular processes other than cytokinesis (for which there has been 3 additional reports [32-34]). Ponti *et al.* used blebbistatin (enantiomeric purity not stated) in a series of live-imaging and small molecule-based experiments [35]. These studies demonstrated that blebbistatin reduced F-actin flow in the lamella actin network, but not in the distinct lamellipodium actin network. This result is consistent with actin-dependent lamella expansion and actomyosin contraction being essential for plasma membrane protrusion and hence cellular locomotion, whilst lamellipodium protrusion is not. Experiments conducted by Grinnell and co-workers [36] using blebbistatin (enantiomeric purity not stated) suggest that platelet-derived growth factor stimulates floating fibroblast-collagen matrix contraction in a myosin-II dependent mechanism. The matrix, which serves as a model for connective tissue, was also able to contract when stimulated with lysophosphatidic acid in the presence of blebbistatin, suggesting that a different motor protein is involved in this case. Blebbistatin has also been used to study a number of other putative myosin II-dependent processes in tissue. These studies have identified a role for myosin II in the formation of apical F-actin rings and disassembly of the epithelial AJC (NMIIA, using (*S*)-(-)-blebbistatin (**2**)) [29], the severing-induced retraction of axons (using blebbistatin of unstated enantiomeric purity) [20] and the cellular migration of pancreatic adenocarcinoma (using blebbistatin of unstated enantiomeric purity) [21].

A significant advantage of (*S*)-(-)-blebbistatin (**2**) as a molecular tool is that it inhibits myosin II directly. Previous to the discovery of **2**, small molecule inhibitors of the kinases that activate myosin II by phosphorylation of its regulatory light chain (MLC) had to be employed. Incomplete inhibition of the kinase activity and constitutive levels of MLC phosphorylation are significant problems with this kinase inhibitor approach [36].

LIMITATIONS OF (*S*)-(-)-BLEBBISTATIN (2) AS A MOLECULAR TOOL

A common approach to studying protein distribution (and by implication function) within a cell is to use fluorescence microscopy techniques on live cells. These experiments often rely on green fluorescent protein (GFP) fusions of the protein under investigation [37].

To visualize the fusion proteins the cells are irradiated with light of a wavelength of 420-490 nm (488 nm in confocal microscopy applications). The light emitted from the protein is collected using pass filters with a typical wavelength range of between 520 and 570 nm. However, (*S*)-(-)-blebbistatin (2) is itself fluorescent [38] limiting its use in experiments of this type. This observation was rationalized by fluorescence emission spectroscopy, which demonstrated that excitation at 440 nm resulted in significant emission by 2 in the GFP emission wavelength range [16]. To facilitate experiments that combine the need for myosin II inhibition with fluorescence imaging of live cells, it is necessary either to use an alternative fluorescent protein (e.g. red fluorescent protein) or to identify an analogue of (*S*)-(-)-blebbistatin (2) with optimized fluorescence emission properties. This serves as another example of the impact that the synthetic chemist can have in chemical genetics research. It was proposed that the addition of a nitro group to the chromophore of 2 would modify its fluorescence properties, although the addition of this functional group must be achieved without loss of biological activity. Synthesis and biological testing of analogues of (*S*)-(-)-blebbistatin (2) had previously demonstrated that substitution at C-7 could be tolerated (Lucas-Lopez *et al.*, unpublished results). Therefore, a sample of (*S*)-(-)-7-nitro-blebbistatin (12) was prepared utilizing our synthetic methodology [16]. Analysis of the fluorescence properties of 12 showed the expected reduction in fluorescence emission in the GFP wavelength range compared with 2. Biochemical assays using 12 showed that it inhibited nonmuscle myosin IIA ATPase activity with an IC_{50} of 28 μM.

Recent microscopy-based studies have identified a further limitation of (±)-blebbistatin [39, 40]. It was shown that prolonged exposure to filtered light (450-490 nm) results in degradation of (±)-blebbistatin to an unidentified non-inhibitory product via cytotoxic intermediates. This degradation was further investigated [16] and found to result from exposure to light in a narrow range around 436 nm.

Patterson, S. *et al.*

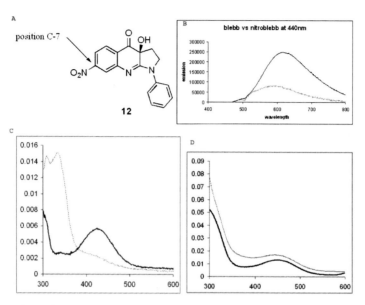

Figure 8. An analogue of (*S*)-(-)-blebbistatin (**2**) with optimized fluorescence properties.

We have recently shown that (*S*)-(-)-7-nitro-blebbistatin (**12**) is stable to prolonged irradiation in this wavelength range. This observation coupled with its reduced fluorescence and retained biological activity suggests that **12** could be a viable alternative to (*S*)-(-)-blebbistatin (**2**) for fluorescence imaging experiments with live cells.

FUTURE STUDIES

Biochemical experiments have shown that (±)-blebbistatin inhibits NMIIA ATPase activity by inhibiting the release of inorganic phosphate from the myosin/ADP/Pi complex [17,41]. This is confirmed and extended in the recently reported structure of (*S*)-(-)-blebbistatin (**2**) bound to the motor domain of *Dictyostelium discoideum* myosin II (Fig. 9) [42]. **2** Binds in a hydrophobic pocket at the apex of a large cleft present in the motor domain close to the γ-phosphate-binding pocket.

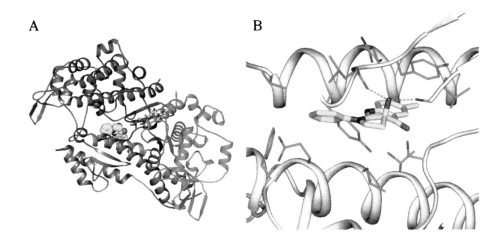

Figure 9. A) The structure of (*S*)-(-)-blebbistatin (**2**) bound to the MgADP-vanadate complex of *D. discoideum* myosin II. **B)** The blebbistatin-myosin binding site. Selected amino acids that interact with (*S*)-(-)-blebbistatin (**2**) are shown in green as their stick representations. For further details see reference [42].

This study opens the way for the rational design of more potent and specific inhibitors of the myosin II subclasses. Through the use of modelling studies, it should also be possible to enhance the discovery of inhibitors of other myosin classes based on the (*S*)-(-)-blebbistain (**2**) core structure or other structures. An alternative approach to the discovery of more potent or more selective myosin II inhibitors would be to perform HTS using libraries of (*S*)-(-)-blebbistatin (**2**) analogues. Whilst the preparation of targeted libraries usually requires a significant amount of chemistry development, the advances in high-throughout synthesis and purification technology help. Our approach to this synthetic challenge involves the development of a solid supported synthesis of (*S*)-(-)-blebbistatin (**2**) analogues using a polymer supported oxidizing reagent (Blum and Westwood, unpublished results) (Fig. 10A) and the use of parallel synthesis/purification technology to prepare small focused collections (Fig. 10B-D) (Westwood, Blum and Lucas-Lopez, unpublished results). These focused libraries concentrate on incorporating substituents in positions that are known not to disrupt NMII ATPase inhibitory activity [6], (Lucas-Lopez *et al.*, unpublished results).

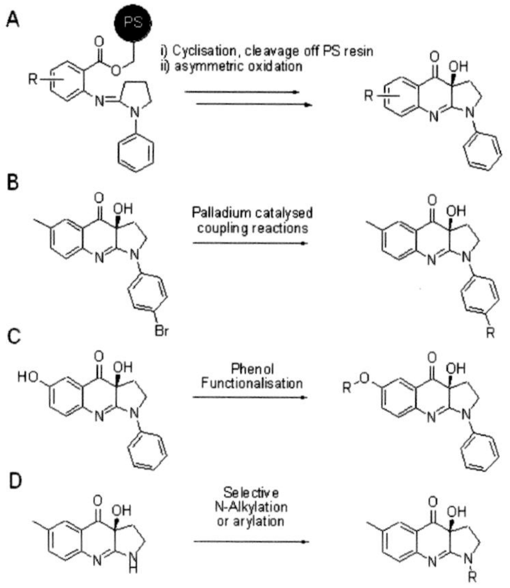

Figure 10. Current strategies for preparing targeted libraries. PS = polystyrene resin.

CONCLUSION

This article has reviewed the discovery, characterization, optimization and use of the novel small molecule tool, (S)-(-)-blebbistatin (2). It provides an overview of the challenges inherent in using chemical genetics to dissect biological mechanisms. This approach is multidisciplinary requiring expertize in cell biology, biochemistry, structural biology as well as computational and synthetic chemistry. Chemical genetics continues to have a significant impact on post-genomic science.

NOTE ADDED IN PROOF

During the editing of this manuscript several further articles describing the use of blebbistatin have appeared [43-52].

ACKNOWLEDGEMENTS

The majority of the research discussed in this article was carried out at the Institute of Chemistry and Cell Biology, Harvard Medical School, U.S.A. (A. F. Straight, T. J. Mitchison, N .J. Westwood); the Laboratory of Molecular Cardiology, National Heart, Lung and Blood Institute, National Institutes of Health, U.S.A. (J. R. Sellers, J. Toth) and the School of Chemistry, University of St Andrews (N. J. Westwood, C. Lucas-Lopez, S. Patterson, T. Blum, A. Slawin). We would like to thank Professor Ivan Rayment, Department of Biochemistry, University of Wisconsin for helpful discussions and the original versions of the images included in Fig. 9. We would also like to acknowledge the Beilstein Institut for the opportunity to take part in their 2004 symposium entitled "The Chemical Theatre of Biological Systems".

REFERENCES

[1]   Begley, T.J., Samson, L.D. (2004) Network responses to DNA damaging agents. *DNA Repair* **3**:1123-1132.

[2]   Heath, J.R., Phelps, M.E., Hood, L. (2003) NanoSystems biology. *Mol. Imaging Biol.* **5**(5):312-325.

[3]   Politz, J.C. (1999) Use of caged fluorochromes to track macromolecular movement in living cells. *Trends Cell Biol.* **9**(7):284-287.

[4]   Mayer, T.U., Kapoor, T.M., Haggarty, S.J., King, R.W., Schreiber, S.L., Mitchison, T.J. (1999) Small-molecule inhibitor of mitotic spindle bipolarity identified in a phenotype-based screen. *Science* **286**(5441):913-914.

[5]   Westwood, N.J. (2004) Chemical Genetics: how does it function? *Phil. Trans. R. Soc. Lond. A* **362**: 761-2774 and references therein.

[6]   Straight, A.F., Cheung, A., Limouze, J., Chen, I., Westwood, N.J., Sellers, J.R., Mitchison, T.J. (2003) Dissecting temporal and spatial control of cytokinesis with a myosin II inhibitor. *Science* **299**:1743-1747.

[7]   Peterson, J.R., Mitchison, T.J. (2002) Small molecules, big impact: a history of chemical inhibitors of the cytoskeleton. *Chem. Biol.* **9**:1275-1285.

[8]   Coue, M., Brenner, S.L., Spector, I, Korn E.D. (1987) Inhibition of actin polymerisation by latrunculin A. *FEBS Lett.* **213**:316-318.

[9]   Mabuchi, I., Okuno, M. (1977) The effect of myosin antibodies on starfish blastomeres. *J. Cell Biol.* **74**:251-263.

[10]   Ostap, E.M. (2002) 2,3-Butanedione monoxime (BDM) as a myosin inhibitor. *J. Muscle Res. Cell Motil.* **23**:305-308.

[11]   Higuchi, H., Takemori, S. (1989) Butanedione monoxime suppresses contraction and ATPase activity of rabbit skeletal muscle. *J. Biochem.* **105**:638-643.

[12]   Cheung, A., Dantzig, J.A., Hollingworth, S., Baylot, S.M., Goldman, Y.E., Mitchison, T.J., Straight, A.F. (2002) A small-molecule inhibitor of skeletal muscle myosin II. *Nature Cell Biol.* **4**:83-88.

[13]   Straight, A.F., Field, C.M., Mitchison, T.J. (2005) Anillin binds nonmuscle myosin II and regulates the contractile ring. *Mol. Biol. Cell* **16**(1):193-201 and references therein.

[14]   Henri, J.F. George, B.A., John, K.N., Austen, M.A. (2001) WPO patent #WO0121796.

[15]   Talaga, P. (2004) Compound decomposition: a new drug discovery tool? *Drug Discov. Today* **9**(2):51-53.

[16]   Lucas-Lopez, C., Patterson, S., Blum, T., Straight, A.F., Toth, J., Slawin, A.M.Z., Mitchison, T.J., Sellers, J.R., Westwood, N.J. (2005) Absolute stereochemical assignment of the small molecule tool, (-)-blebbistatin. *Eur. J. Org. Chem.* in press.

[17]   Kovacs, M., Toth, J., Hetenyi, C., Malnasi-Csizmadia, A., Sellers, J.R. (2004) Mechanism of blebbistatin inhibition of myosin II. *J. Biol. Chem.* **279**:35557-35563.

[18]   Davis, F.A., Chen, B.-C. (1992) Asymmetric hydroxylation of enolates with *N*-sulfonyloxaziridines. *Chem. Rev.* **92**:919-934.

[19]   Davis, F.A., Weismiller, M.C., Murphy, C.K., Reddy, T., Chen, B.-C. (1992) Chemistry of oxaziridines. 18. Synthesis and enantioselective oxidations of the [(8,8-dihalocamphoryl)sulfonyl]oxaziridines. *J. Org. Chem.* **57**:7274-7285.

[20]   Gallo, G. (2004) Myosin II activity is required for severing-induced axon retraction in vitro. *Exp. Neurol.* **189**:112-121.

[21]   Duxbury, M.S., Ashley, S.W., Whang, E.E. (2004) Inhibition of pancreatic adenocarcinoma cellular invasiveness by blebbistatin: a novel myosin II inhibitor. *Biochem. Biophys. Res. Commun.* **313**:992-997.

[22]   Graves, P.R., Kwiek, J.J., Fadden, P., Ray, R., Hardeman, K., Coley, A.M., Foley, M., Haystead, T.A. (2002) Discovery of novel targets of quinolone drugs in the human purine binding proteome. *Mol. Pharmacol.* **62**(6):1364-1372.

[23]   Sellers, J.R. (2000) Myosins: a diverse superfamily. *Biochim. Biophys. Acta* **1496**:3-22.

[24]   Berg, J.S., Powell, B.C., Cheney, R.E. (2001) A millennial myosin census. *Mol. Biol. Cell* **12**:780-794.

[25]   Hodge, T., Cope, M.J.T.V. (2000) A myosin family tree. *J. Cell Sci.* **113**:3353-3354.

[26]   Golomb, E., Ma, X., Jana, S.S., Preston, Y.A., Kawamoto, S., Shoham, N.G., Goldin, E., Conti, M.A., Sellers, J.R., Adelstein R.S. (2004) Identification and characterisation of nonmuscle myosin II-C, a new member of the myosin II family. *J. Biol. Chem.* **279**:2800-2808.

[27]   Davies S.P., Reddy, H., Caivano, M., Cohen, P. (2000) Specificity and mechanism of action of some commonly used protein kinase inhibitors. *Biochem. J.* **351**(1):95-105.

[28]   Limouze, J., Straight, A.F., Mitchison, T., Sellers, J.R. (2004) Specificity of Blebbistatin, an inhibitor of myosin II. *J. Muscle Res. Cell Motil.* **25**:337-341.

[29]   Ivanov, A.I., McCall, I.C., Parkos, C.A., Nusrat, A (2004) Role for actin filament turnover and a myosin II motor in cytoskeletal-driven disassembly of the epithelial apical junction complex. *Mol. Biol. Cell.* **15**:2639-2651.

[30]   Meissner, M., Schlüter, D., Soldati, D. (2002) Role of Toxoplasma gondii myosin A in powering parasite gliding and host cell invasion. *Science* **298**:837-840 and references therein.

[31]   Carey, K.L., Westwood, N.J., Mitchison, T.J., Ward, G.E. (2004) A small-molecule approach to study invasive mechanisms of *Toxoplasma gondii*. *Proc. Natl. Acad. Sci. USA* **101**:7433-7438.

[32]   Rosenblatt, J., Cramer, L.P., Baum, B., McGee, K.M. (2004) Myosin II-dependent cortical movement is required for centrosome separation and positioning during mitotic spindle assembly. *Cell* **117**:361-372.

[33]   Uetake, Y., Sluder, G. (2004) Cell cycle progression after cleavage failure: mammalian somatic cells do not possess a "tetraploidy checkpoint." *J. Cell Biol.* **165**:609-615.

[34]   Pielak, R.M., Gaysinskaya, V.A., Cohen, W.D. (2004) Formation and function of the polar body contractile ring in Spinsula. *Devel. Biol.* **269**:421-432.

[35]   Ponti, A., Machacek, M., Gupton, S.L., Waterman-Storer, C.M., Danuser, G. (2004) Two distinct actin networks drive the protrusion of migrating cells. *Science* **305**:1782-1786.

[36]   Abe, M., Ho, C-H., Kamm, K.E. Grinnell, F. (2004) Different molecular motors mediate platelet-derived growth factor and lysophosphatidic acid-stimulated floating collagen matrix contraction. *J. Biol. Chem.* **278**:47707-47712.

[37]   Ehrhardt, D. (2003) GFP technology for live cell imaging. *Curr. Opin. Plant. Biol.* **6**:622-628.

[38]   Sakamoto, T., Limouze, J., Combs, C.A., Straight, A.F., Sellers, J.R. (2005) Blebbistatin, a myosin II inhibitor, is photoinactivated by blue light. *Biochemistry* **44**(2):584-588.

[39]   Kolega, J. (2004) Phototoxicity and photoinactivation of blebbistatin in UV and visible light. *Biochem. Biophys. Res. Commun.* **320**:1020-1025.

[40]   Sakamoto, T., Limouze, J., Combs, C.A., Straight, A.F., Sellers, J.R. (2005) Blebbistatin, a myosin II inhibitor, is photoinactivated by blue light. *Biochemistry* **44**:584-588.

[41]   Ramamurthy, B., Yengo, C.M., Straight, A.F., Mitchison, T.J., Sweeney, H.L. (2004) Kinetic mechanism of blebbistatin inhibition of nonmuscle myosin IIB. *Biochemistry* **43**:14832-14839.

[42]   Allingham, J.S., Smith, R., Rayment, I. (2005) The structural basis of blebbistatin inhibition and specificity for myosin II. *Nature Struct. Mol. Biol.* Advance online publication (doi:10.1038/nsmb908).

[43]   Yoshizaki, H., Ohba, Y., Parrini, M.C., Dulyaninova, N.G., Bresnick, A.R., Mochizuki, N., Matsuda, M. (2004) Cell type-specific regulation of RhoA activity during cytokinesis. *J. Biol. Chem.* **279**:44756-44762.

[44]   Griffin, M.A., Sen, S., Sweeney, H.L., Discher, D.E. (2004) Adhesion-contractile balance in myocyte differentiation. *J. Cell Sci.* **117**:5855-5863.

[45]   Bastian, P., Lang, K., Niggemann, B., Zaenker, K.S., Entschladen F. (2005) Myosin regulation in the migration of tumor cells and leukocytes within a three-dimensional collagen matrix. *Cell. Mol. Life Sci.* **62**:65-76.

[46]   Shu, S., Liu, X., Korn, E.D. (2005) Blebbistatin and blebbistatin-inactivated myosin II inhibit myosin II-dependent processes in *Dictyostelium. Proc. Nat. Acad. Sci. USA* **102**:1472-1477.

[47]   Ekman, M., Fagher, K., Wede, M., Stakeberg, K., Arner A. (2005) Decreased phosphatase activity, increased Ca^{2+} sensitivity, and myosin light chain phosphorylation in urinary bladder smooth muscle of newborn mice. *J. Gen. Physiol.* **125**:187-196.

[48]   Wong, C., Stearns, T. (2005) Mammalian cells lack checkpoints for tetraploidy, aberrant centrosome number, and cytokinesis failure *BMC. Cell. Biol.* Advance online publication (doi:10.1186/1471-2121-6-6).

[49]   Wilkinson, S., Paterson, H.F., Marshall, C.J. (2005) Cdc42-MRCK and Rho-ROCK signalling cooperate in myosin phosphorylation and cell invasion. *Nature Cell. Biol.* **7**:255-261.

[50]   Di Ciano-Oliveira, C., Lodyga, M., Fan, L., Szaszi, K., Hosoya, H., Rotstein, O.D., Kapus, A. (2005) Is myosin light chain phosphorylation a regulatory signal for the osmotic activation of the Na^+-K^+-$2Cl^-$ cotransporter. *Am. J. Physiol. Cell Physiol.* Advance online publication (doi:10.1152/ajpcell.00631.2004).

[51]   Xia, D., Stull, J.T., Kamm, K.E. (2005) Myosin phosphatase targeting subunit 1 affects cell migration by regulating myosin phosphorylation and actin assembly. *Exp. Cell Res.* **304**:506-517.

[52]   Ivanov, A.I., Hunt, D., Utech, M., Nusrat, A., Parkos, C.A. (2005) Differential roles for actin polymerisation and a myosin II motor in assembly of the epithelial apical junction complex. *Mol. Biol. Cell.* Advance online publication.

MULTIDIMENSIONAL EXPLORATION INTO BIOCHEMICAL PATHWAYS

JOHANN GASTEIGER[*1,2], MARTIN REITZ[1] AND OLIVER SACHER[2]

[1]Computer-Chemie-Centrum and Institute of Organic Chemistry, University of Erlangen-Nuremberg, Naegelsbachstraße 25, 91052 Erlangen, Germany

[2]Molecular Networks GmbH, Naegelsbachstraße 25, 91052 Erlangen, Germany

E-Mail: *gasteiger@chemie.uni-erlangen.de

Received: 17th December 2004 / Published: 22nd July 2005

ABSTRACT

The famous Biochemical Pathways wall chart has been converted into a reaction database. The web based retrieval system C@ROL has been interfaced to this BioPath database providing a wide variety of search methods for chemical structures, enzymes, and reactions that can allow one to explore the endogenous metabolism of different species. The database has been made accessible on the internet at:
http://www2.chemie.uni-erlangen.de/services/biopath/index.html and
http://www.mol-net.de/databases/biopath.html.

It is shown how the information in this database can be used to explore enzyme inhibitors as transition state mimics. Furthermore, it is shown how the classification of biochemical reactions based on physico-chemical effects at the reaction site, corresponds with the classification of enzymes by the EC code.

INTRODUCTION

Massive efforts have gone into the deciphering of the human genome. With this goal achieved an important milestone has been reached. However, it was clear from the very beginning that this can only be the start of our understanding of the processes that keep us alive and which might go wrong thus causing diseases.

Thus, attention has shifted to proteomics, the study of the proteins that regulate these processes. Furthermore, increasing research is devoted to metabolomics, the study of the compounds that are produced in living organisms and the way in which they are produced.

Even before the advent of genomics a large body of evidence had been accumulated on the compounds and reactions occurring in our bodies, the biochemical reactions. The essential results of all this work have been compiled quite beautifully in the famous Biochemical Pathways wall chart, produced by G. Michal and coworkers, and initially distributed by Boehringer Mannheim, and now by Roche [1]. These biochemical pathways are outlined in greater detail in an Atlas on Biochemical Pathways [2].

The task now, is to bring these two avenues of investigation, genomics and proteomics on one hand, and biochemical pathways on the other, closer together to link these sources of information. On one side, with genomics and proteomics we have huge amounts of data that have to be processed by bioinformatics methods. And on the side of biochemical pathways we have quite detailed information on small molecules and their interconversions that ask for a more thorough analysis. This is where chemoinformatics can come in, a field with a long history of developing methods for the representation, manipulation and analysis of chemical structures and reactions, which has matured to a discipline providing powerful methods [3,4].

BIOCHEMICAL PATHWAY DATABASES

The Biochemical Pathways wall chart contains a cornucopia of information in a highly condensed manner. This makes it difficult to locate individual compounds, reactions, or enzymes. Figure 1 shows a small selection of this wall chart which by the very fact of its outline of reaction arrows emphasizes the problem: we have a highly connected network of reactions that emerge and zero into compounds that sometimes participate in quite as many reactions.

In essence, biochemical pathways form a high-dimensional space that had to be projected into two dimensions to produce the poster. The task is then to exploit the full high-dimensionality of biochemical pathways and allow it to be explored. This is where chemoinformatic methods can come in to allow searches for structures, substructures, reactions and enzymes. To achieve this task, the information on the Biochemical Pathways wall chart had to be stored in a database of structures and reactions.

Multidimensional Exploration into Biochemical Pathways

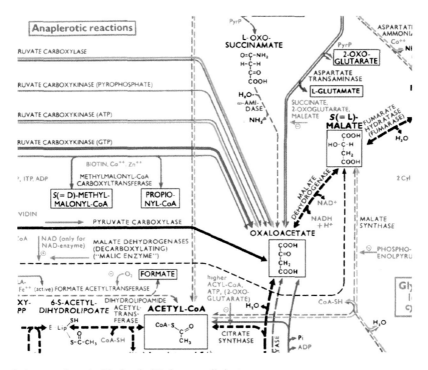

Figure 1. A cut out from the Biochemical Pathways wall chart.

This is exactly what we did:

> · metabolites, coenzymes, and regulators were stored by their name and as connection tables;
>
> · enzymes were stored by name and EC code;
>
> · chemical reactions were indicated by their starting materials and products as well as the enzymes and coenzymes involved. Furthermore, it was indicated whether a reaction occurs as a general pathway, only in animals, in higher plants, or in unicellular organisms. Above all, the bonds broken and made in a reaction were marked.

The representation of chemical structures on the atomic level indicating all atoms and bonds in a molecule in the form of a connection table, allows one to use the full arsenal of chemical database searching methods such as full structure and substructure searching. At the outset of our work no database on biochemical reactions containing structures in the form of connection tables was available.

In the meantime several such databases have appeared such as KEGG [3] and the collection of databases on the BioCyc [4] web page. However, none of these databases have marked the reaction centre, an essential feature for correct reaction searching, as we will see later. This is a unique feature of our BioPath database.

At present, the BioPath database contains about 1500 structures and about 2200 reactions. Having a database with chemical structures in the form of connection tables allows one to enrich the database by data generated by chemoinformatic methods. Thus, we have sent all the molecules contained in the BioPath database through our 3D structure generator CORINA [5,6] and the program ROTATE [7,8] that generates multiple conformations. For each metabolite, coenzyme, and regulator, 3D molecular models for an ensemble of conformations have been stored in the BioPath database, allowing 3D substructure searching. Figure 2 shows a molecular model of coenzyme A as obtained by CORINA.

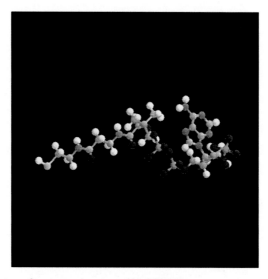

Figure 2. The 3D structure of coenzyme A generated by CORINA.

Furthermore, a variety of links have been added to the BioPath database, particularly those to other bioinformatics databases such as BRENDA [9] or those on the ExPASy server [10].

SEARCHING IN THE BIOPATH DATABASE

In order to exploit the rich information content in the BioPath database we have developed the web based C@AROL retrieval system [11], which allows searches either on chemical structure information or on chemical reaction information.

Chemical structures can be searched by:

> ·name or name fragments
>
> ·full structure
>
> ·substructure
>
> ·structure similarity
>
> ·3D substructure

Enzymes can be searched by:

> ·name
>
> ·EC code

Chemical reactions can be searched by:

> ·The structure of the starting materials or products
>
> ·Enzyme name or EC code
>
> ·Chemical transformation

Clearly, chemical transformation searching is a very important feature because it allows one to search for the essential transformation of a substrate invoked by an enzyme. For such transformation searching, it is essential that the atoms and bonds directly participating in a reaction are marked. In reaction searching it does not suffice to specify the functional groups in the starting materials and in the interconversion of products in which one is interested. Rather, one has also to specify the mapping of the atoms and bonds in the functional groups. Thus, if one is interested in the oxidation of primary alcohols to aldehydes and only specifies that the starting materials should contain an alcohol group and the product should contain an aldehyde group one would obtain the reaction shown at the bottom of Fig. 3. However, in reality this reaction involves the phosphorylation of glyceraldehyde to glyceraldehyde-3-phosphate as a hit, for the starting material having an alcohol function and the product an aldehyde group.

Only if one is indicating the bonds broken and made in a reaction and requires the atoms of the bonds broken to be mapped onto the atoms of the bonds made in the reaction will it be realized that this reaction is a phosphorylation and not an oxidation of an alcohol to an aldehyde.

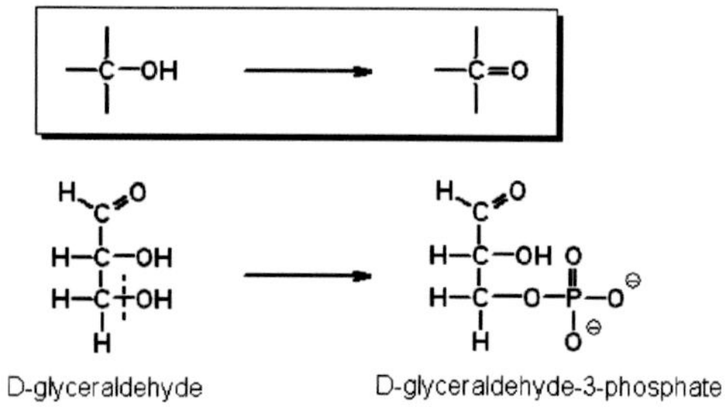

Figure 3. The phosphorylation of glyceraldehydes to glyceraldehyde-3-phosphate (bottom) obtained as a result of the substructure search shown on top.

We do not have sufficient space here to outline all the rich and diverse search possibilities in the BioPath database. A more extensive presentation has recently been published [12]. Only a few examples are presented here. The reader is encouraged to explore the BioPath database by her/himself as it has been made accessible together with the C@ROL retrieval system on the web at:

http://www2.chemie.uni-erlangen.de/services/biopath/index.html and

http://www.mol-net.de/databases/biopath.html

Structure and Substructure Searching

The structure editor JME [13] has been integrated into C@ROL allowing easy graphical specification of structure or substructure queries. Figure 4 shows the query for a substructure search. Here the substructure tetrahydropyrane has been input. The search resulted in 83 hits for this example.

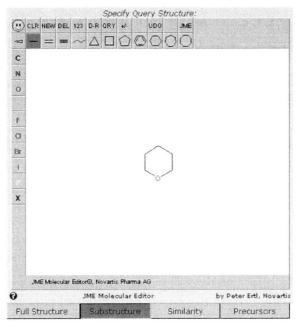

Figure 4. Input of a substructure search into C@ROL via JME.

Reaction Searching

Searching by enzyme

Inputting the EC number 3.1.3.3 provided two hits, one reaction being indicated in Fig. 5, the reaction catalysed by phosphoserine phosphatase.

Following the link in the ExPASy server contained in the BioPath database provides the image shown in Fig. 6.

Thus, you can see in which environment this reaction is embedded. Incidentally, this also shows one of the advantages of the BioPath database that clearly highlights this reaction, whereas it is hard to extract from the image in Fig. 6. The markings are ours and are not contained in the image on the ExPASy server.

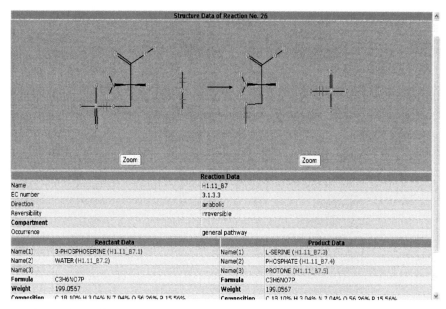

Figure 5. The reaction catalysed by the enzyme phosphoserine phosphatase (EC 3.1.3.3).

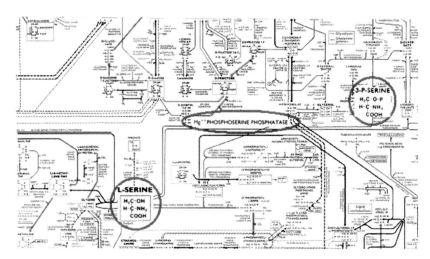

Figure 6. The reaction of phosphoserine phosphatase (EC 3.1.3.3) depicted on the Biochemical Pathways wall chart.

Searching by transformation

Oxidations are one of the most important reactions in metabolism, on one hand providing energy, and, on the other hand, transforming organic compounds into more water-soluble species, an important step in the excretion of xenobiotics. Figure 7 shows the query for searching for oxidations of C-H bonds to C-OH bonds.

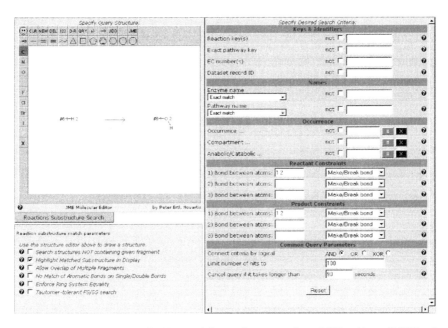

Figure 7. Reaction centre search in C@ROL: search for reactions converting a C-H bond into a C-OH bond.

This query provided 97 reactions as hit underlining the importance of this reaction type. Figure 8 shows one example from this hit list, the oxidation of leukotriene to 20-hydroxy-leukotriene.

Gasteiger, J. *et al.*

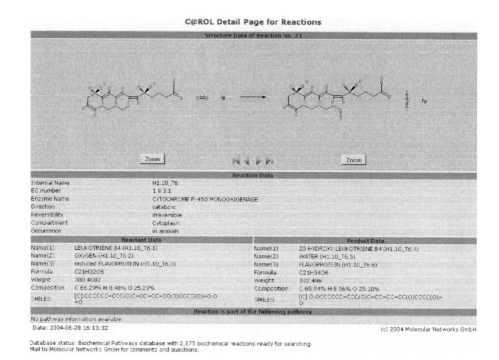

Figure 8. One of 97 hits of reactions of the query of Fig. 7 converting a C-H bond into a C-OH bond.

APPLICATIONS

We consider the BioPath database as an important milestone in the understanding of the processes that keep living species alive. It is an information source at the interface of bioinformatics and chemoinformatics. Bioinformatics is concerned with the expression of enzymes by genes, whereas now we can use chemoinformatics methods to study the reactions that are controlled by the enzymes, on the very detailed level of individual atoms and bonds in the molecules. We are just at the very beginning of exploiting this rich source of information. Here, two examples are given of the use of the BioPath database to explore details of enzyme action.

Inhibitors as Transition State Analogues

Nearly 60 years ago Linus Pauling emphasized that enzymes are complementary in structure to the activated complexes of the reactions that they catalyse.

In other words, enzymes should bind the transition state of a reaction with far greater affinity than the substrate or the products. As a case in point we have analysed the reaction catalysed by AMP deaminase (EC 3.5.4.6) (Fig. 9) converting adenosine-monophosphate (AMP) into inosine-monophosphate (IMP).

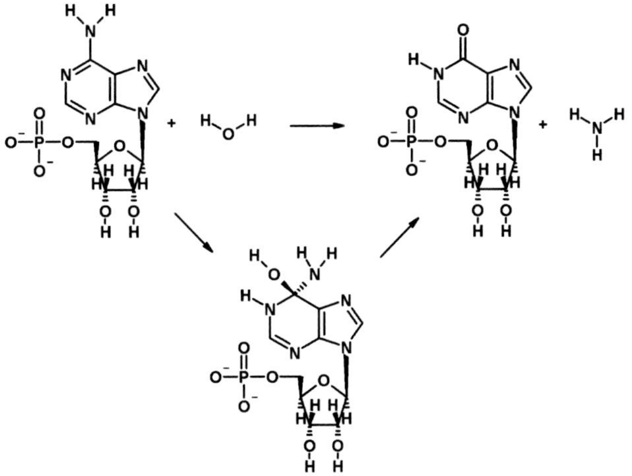

Figure 9. The reaction of AMP deaminase, converting AMP into IMP. The reaction intermediate of this reaction is shown at the bottom.

The reaction quite certainly proceeds first through the addition of a water molecule to produce the intermediate, also shown in Fig. 9; the transition state of this reaction being close in geometry to this intermediate. Coformycin (Fig. 10) is an inhibitor of this enzyme.

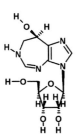

Figure 10. An inhibitor of AMP deaminase: coformycin.

Following Pauling one would conjecture that this inhibitor should be closer in geometry to the intermediate than to AMP or IMP. To test this hypothesis, the inhibitor coformycin was superimposed with AMP, IMP, and the intermediate, respectively. This was done with the aid of the program GAMMA (Genetic Algorithm for Multiple Molecule Alignment) [14]. Figures 11-13 show the three superimpositions.

As can be seen, the geometry of the inhibitor fits best with the geometry of the intermediate, in particular, the two OH-groups are almost perfectly aligned. Apparently, the orientation of this incoming OH-group - which one knows to be coordinated to a Zn^{2+} ion - plays a crucial role in this reaction.

Figure 11. 3D superimposition of AMP with the inhibitor coformycin.

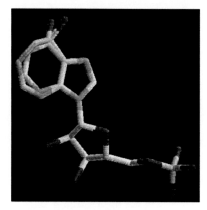

Figure 12. 3D superimposition of IMP with the inhibitor coformycin.

Figure 13. 3D superimposition of the AMP deaminase reaction intermediate with the inhibitor coformycin. The OH groups, which are crucial for proper binding, are very well aligned.

Various other enzyme-catalysed reactions have been studied along similar lines by comparison of the geometry of inhibitors of enzymes [15], with the transition states or intermediates of the reaction catalysed by these enzymes.

All these studies support this hypothesis that an inhibitor of an enzyme should be particularly similar in geometry to the transition state of the reaction catalysed by this enzyme.

Classification of Enzymes vs. Classification of Reactions

Enzymes are classified by the widely accepted EC code [16] that builds its classification on a variety of criteria such as reaction patterns, substrates, transferred groups, and acceptor groups. Thus, the EC classification is not quite coherent as, depending on the EC class, the emphasis shifts between different criteria.

Clearly, the most important action of an enzyme is the catalysis of a reaction, an event that breaks and makes bonds. The question therefore is can we build the classification of enzymes on considerations that only take into account the properties of the bonds directly involved in the reaction event? In other words, we want to classify enzyme-catalysed reactions and compare this classification with the EC classification system.

Biochemical reactions are governed by the same kind of physico-chemical effects as more traditional organic reactions, effects that are involved in the discussion of reaction mechanism such as charge distribution, inductive, resonance, or polarizability affect. Some time ago, we developed procedures that allow the calculation of these affects providing values that have shown their importance in modelling reaction mechanisms [17,18].

In order to investigate how reaction classification corresponds with enzyme classification we have chosen reactions that are catalysed by hydrolases. These fall into the EC category 3.1.x.y. The question is then, how will the classification of reactions catalysed by enzymes match classes 3.1.1.y, 3.1.2.y, 3.1.3.y, 3.1.4.y, 3.1.5.y, and 3.1.6.y?

Each bond broken in the reaction with water was characterized by six physico-chemical values, calculated by simple empirical procedures: difference in total charges [19], difference in σ-electronegativities [19], difference in π-electronegativities [20], effective bond polarizabilities [21], and delocalization stabilization of a positive or negative charge. In effect, each reaction is then an event in a six-dimensional space spanned by the above six physico-chemical descriptors as coordinates, with each bond hydrolysed having a specific value for each of these six coordinates.

In order to visualize the distribution of each reaction in this six-dimensional space we projected this space into two dimensions by a Kohonen neural network [22,23]. The dataset contained 49 reactions which were mapped into a 8x8 Kohonen network. Figure 14 shows the map thus obtained.

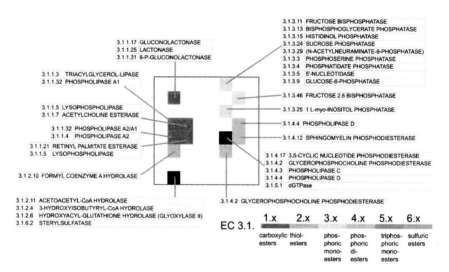

Figure 14. Kohonen map showing the classification of reactions catalysed by enzymes of EC class 3.1.x.y (49 reactions in dataset).

As can be seen, reactions catalysed by enzymes of different subclasses of the EC code 3.1. are mapped, by and large, into different, well separated areas of the two dimensional map. There are only two squares (neurons) where reactions belonging to different subclasses of EC 3.1. are mapped. Reactions catalysed by enzymes of subclass 3.1.4. and one catalysed by an enzyme of subclass 3.1.5. are mapped in one square. However, it should be realized, that the enzyme EC 3.1.5.1, dGTPase, involves the hydrolysis of a phosphate much like the enzymes of EC 3.1.4.y. Thus, indeed these reactions have much in common, and somehow the EC classification artificially separates them. In the other case, reactions catalysed by enzymes of EC 3.1.2. are mapped, all invoking hydrolysis of thioesters, and one reaction catalysed by EC 3.1.6.2, sterylsulfatase.

Similar results have been obtained in investigations of other EC classes. This leads us to emphasize that the classification of reactions based on physico-chemical affects corresponds to a large extent to the EC classification system.

However, the classification based on reactions provides chemically sounder results and also shows finer details in the reactions catalysed by the enzymes.

SUMMARY

The BioPath database constitutes a rich source of information on the structures and reactions involved in the endogenous metabolism. The web based retrieval system C@ROL provides a host of search methods that allow the exploration of these biochemical pathways. We are just at the beginning of discovering new insights into the all-important reactions that keep living species alive. Two examples were given here. The three-dimensional structures of the intermediates in biochemical reactions provide insight into the geometric situations in the binding pocket of enzymes, information that can be used to search for inhibitors of enzymes. In the other application we have shown that the classification of enzyme catalysed reactions based on the physico-chemical affects operating on the bonds directly involved in the reactions corresponds by and large with the EC code classification system. However, it constitutes a chemically more meaningful classification and provides new insights into enzyme reactions.

ACKNOWLEDGEMENTS

We appreciate assistance in the construction of the BioPath database and the analysis of its contents through projects funded by the Bundesministerium fuer Bildung und Forschung (BMBF projects no. 08 C 5850 0, 08 C 5879, 031U112D, 031U212D, 031U112A, and 031U212A). We appreciate the initiation of the Biochemical Pathways project by Spektrum Akademischer Verlag and the collaboration with Prof. Guido Moerkotte and Dr Carl-Christian Kanne, University of Mannheim, Germany in establishing our data scheme. Dr Wolf-Dietrich Ihlenfeldt, at that time at the CCC, provided important contributions to this project, most notably the CACTVS system. Discussions with Dr Gerhard Michal were always stimulating. We are indebted to a number of students who carefully input the structures and reactions into the BioPath database. The BFAM project, initiated by Prof. Hans-Werner Mewes allowed us to continue the work on the BioPath database.

REFERENCES

[1]   Michal, G. (Ed.) (1993) Biochemical Pathways Wall Chart. Boehringer Mannheim, (now Roche), Germany.

[2]   Michal, G. (Ed.) (1999) Biochemical Pathways - Biochemical Atlas. Spektrum Akademischer Verlag, Heidelberg, Germany.

[3]   Goto, S., Okuno, Y., Hattori, M., Nishioka, T., Kanekisa, H. (2002) LIGAND:database of chemical compounds and reactions in biological pathways. *Nucl. Acids Res.* **30**(1): 402-404.

[4]   BioCyc Database Collection, http://www.biocyc.org.

[5]   Sadowski, J., Gasteiger, J., Klebe, G. (1994) Comparison of automatic three-dimensional model builders using 639 X-ray structures. *J. Chem. Inf. Comput. Sci.* **34**:1000-1008.

[6]   CORINA can be tested online at http://www2.chemie.uni-erlangen.de/software/corina/free_struct.html and is available from Molecular Networks GmbH, Germany, info@mol-net.de, http://www.mol-net.de.

[7]   Schwab, C.H. (2003) In: *Handbook of Chemoinformatics - From Data to Knowledge.* (Gasteiger, J., Ed.), pp.262-301. Wiley-VCH, Weinheim, Germany.

[8]   ROTATE is available from Molecular Networks GmbH, Germany, info@mol-net.de, http://www.mol-net.de.

[9]   BRENDA - The Comprehensive Enzyme Information System, http://www.brenda.uni-koeln.de.

[10]   ExPASy Server, University of Geneva, Switzerland, http://www.expasy.org/tools/pathways.

[11]   C@ROL is available from Molecular Networks GmbH, Germany, info@mol-net.de, http://www.mol-net.de.

[12]   Reitz, M., Sacher, O., Tarkhov, A., Trümbach, D., Gasteiger, J. (2004) Enabling the exploration of biochemical pathways. *Org. Biomol. Chem.* **2**(22): 3226-3237.

[13]   JME molecule editor, developed by Ertl, P., Novartis, available from Molinspiration Chemoinformatics, http://www.molinspiration.com.

[14]   Handschuh, S., Wagener, M., Gasteiger, J.J. (1988) Superposition of three-dimensional chemical structures allowing for conformational flexibility by a hybrid method. *Chem. Inf. Comput. Sci.* **38**: 220-232.

[15]   Reitz, M., Gasteiger, J., in preparation.

[16]   *Enzyme Nomenclature* (1992), IUBMB. Academic Press, San Diego, California.

[17]   Gasteiger, J. (2003) Physicochemical effects in the representation of molecular structures for drug designing. *Mini Rev. Med. Chem.* **3**:789-796.

[18]   Gasteiger, J., Marsili, M., Hutchings, M.G., Saller, H., Löw, P., Röse, P., Rafeiner, K. (1990) Models for the representation of knowledge about chemical reactions. *J. Chem. Inf. Comput. Sci.* **30**:467-476.

[19]   Gasteiger, J., Marsili, M. (1980) Iterative partial equalization of orbital electronegativity - a rapid access to atomic charges. *Tetrahedron* **36**: 3219-3228.

[20]   Gasteiger, J., Saller, H. (1985) Calculation of the charge distribution in conjugated systems by a quantification of the resonance concept. *Angew. Chem. Int. Ed. Engl.* **24**: 687-689.

[21]   Gasteiger, J., Hutchings, M. G. (1984) Quantification of effective polarisability. Applications to studies of X-ray photoelectron spectroscopy and alkylamine protonation. *J. Chem. Soc. Perkin*, **2**, 559-564.

[22]   Kohonen, T. (1995) In: *Self Organizing Maps.* (Huang, T. S., Kohonen, T., Schröder, M. R., Eds). Springer Verlag, Berlin.

[23]   Zupan, J., Gasteiger, J. (Eds) (1999) *Neural Networks in Chemistry and Drug Design*, Vol. 2. Wiley-VCH, Weinheim.

Beilstein-Institut   The Chemical Theatre of Biological Systems, May 24th - 28th, 2004, Bozen, Italy

HANDLING EQUILIBRIUM PROCESSES EMBEDDED IN METABOLIC SYSTEMS

ATHEL CORNISH-BOWDEN AND MARÍA LUZ CÁRDENAS

Institut de Biologie Structurale et Microbiologie,
Centre National de la Recherche Scientifique, 31 chemin Joseph-Aiguier, B.P. 71, 13402
Marseille Cedex 20, France

E-Mail: acornish@ibsm.cnrs-mrs.fr

Received: 9th August 2004 / Published: 22nd July 2005

ABSTRACT

Thermodynamic constraints are essential for understanding the behaviour of living systems, but they are far from sufficient, because they allow a wide range of possibilities. Additional constraints imposed by kinetic considerations are often crucial in determining not merely whether a process can occur, but whether it does. Entropies of activation are in principle very useful for analysing experimental properties, but in practice they are rendered almost useless by the impossibility of estimating them accurately from observations spread over a narrow temperature range. Such estimation typically involves extrapolating more than 10-times the range of the data, and involves huge errors. Another common problem in the literature, results from confusion between actual and standard Gibbs energies of reaction: supposedly unfavourable equilibrium constants can suggest that processes necessary for life, such as reduction of sulphate by organisms that use sulphate as a terminal electron acceptor, are impossible; but as long as an organism has efficient mechanisms for maintaining reactant concentrations far from their standard states, a reaction can be driven in either direction regardless of the magnitude of its equilibrium constant. The increasing importance of computer modelling in studying metabolism has now focused attention on the question of how to handle reactions that are essentially irreversible. It has often been assumed that if the reverse reaction is negligible then all the effects of products can be neglected, but that is a potentially serious error: many enzymes are known where product inhibition has important effects on the rate of a reaction that is for practical purposes irreversible.

INTRODUCTION

Living organisms are chemical systems that operate far from thermodynamic equilibrium, and it is therefore difficult to predict their behaviour from simple thermodynamic considerations alone. This does not mean, of course, that metabolic processes do not obey thermodynamic rules: they do, but although thermodynamic constraints limit what is permitted they still allow a wide range of possibilities. More generally, the laws of physics place limits on what is possible in a living system, but they fall far short of defining what a living system is, or even of predicting that life can exist [1]. The difference between the standard Gibbs energy and the Gibbs energy of a particular physiological state, which may be very large, can never be neglected. Furthermore, kinetic constraints are also important, and are often crucial in determining whether a process occurs. In this article we discuss a number of different thermodynamic aspects of biochemistry that are sometimes forgotten or badly understood, though they need proper attention if valid metabolic models are to be created. Most of our points are not new, and some, indeed, date back to the early years of thermodynamics. Accordingly, we prefer to provide a general discussion, with references to more detailed arguments, rather than presenting all of the supporting details here.

We begin by examining why thermodynamics has always been regarded as a difficult subject, suggesting that the problems has its roots in the rather obscure way in which the essential concept of entropy was originally introduced and defined. This has led, on the one hand, to major examples of its misuse, for example in most evocations of the idea of entropy-enthalpy compensation, and, on the other hand, to reluctance to accept perfectly valid concepts such as enzymes that are more effective as catalysts in one direction of reaction than the other. Most serious, from the point of view of metabolic modelling, uncertainty about the proper way to deal with reactions with very large equilibrium constants has led both to valid models that are more complicated than they need to be, and to others that are invalid because they use thermodynamic arguments to reach invalid kinetic conclusions.

ORIGINS OF ENTROPY

The concept of entropy was introduced to thermodynamics by Clausius, who deliberately chose an obscure term for it, wanting a word based on Greek roots that would sound similar to "energy".

In this way he hoped to have a word that would mean the same to everyone regardless of their language, and, as Cooper [2] remarked, he succeeded in this way in finding a word that meant the same to everyone: nothing. From the beginning it proved a very difficult concept for other thermodynamicists, even including such accomplished mathematicians as Kelvin and Maxwell; Kelvin, indeed, despite his own major contributions to the subject, never appreciated the idea of entropy [3]. The difficulties that Clausius created have continued to the present day, with the result that a fundamental idea that is absolutely necessary for understanding the theory of chemical equilibria continues to give trouble, not only to students but also to scientists who need the concept for their work.

ENTROPY-ENTHALPY COMPENSATION

Partitioning temperature effects on reactions into entropy and enthalpy components is very useful for understanding why they behave as they do. A major practical problem arises however, from the fact that the actual numbers are almost impossible to estimate accurately from temperature-dependence measurements over the narrow ranges of temperature typically accessible for biological systems. Calorimetric measurements, as used for example by Freire and co-workers [4], are another matter, of course and are not open to the objections that we shall discuss, but unfortunately they are not the ones most frequently encountered; in the biochemical literature, and much other chemical literature, thermodynamic parameters are nearly always the result of analysing the temperature dependencies of equilibrium or kinetic constants. Even today, reputable journals provide examples of entropy estimates given with a precision of about 10%, even though they come from measurements at four or five temperatures in a narrow range. The entropy estimate is then obtained by extrapolating a small amount of information over more than ten times the range of the data (Fig. 1).

A study of the dependence on temperature of the kinetic parameters of an enzyme is typically restricted to a range from a little above 0°C to no more than 40°C. For various practical reasons the range is often substantially smaller than this. For example, a study of ATPases of various fishes [7], used as an example in a current textbook of biochemical kinetics [8], involved measurements in the range 0-18°C.

In the ordinary units used for Arrhenius plots, this is a $1000/T$ range of 3.44-3.66 K^{-1}, and so the extrapolation needed to estimate the intercept on the ordinate axis, and hence the entropy of activation, extends more than 15 times the range of the data. Despite this, the apparent correlation between entropies and enthalpies of activation of the ATPases of seven different (and in most cases very distantly related) fishes is excellent, with no perceptible deviation from a perfect linear relationship. In our view, the only reasonable interpretation of such a perfect correlation is a statistical artefact caused by the extremely high correlation that inevitably appears between a slope and an intercept estimated from an excessively long extrapolation [9]. In fact, simulated data with mean catalytic activities varying randomly over a 10-fold range and Arrhenius activation energies varying randomly in the range 25-160 kJ/mol still show an excellent, but meaningless, correlation when transformed into a compensation plot [10].

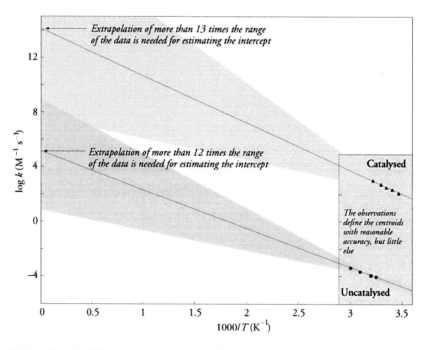

Figure 1. Extrapolation implicit in estimation of entropies from Arrhenius plots. The shaded area is redrawn from Fig. 4 of [5] and shows the temperature dependence of rate constants for uncatalysed and catalysed formation of peptide bonds. The data were said to provide estimates accurate to ±10% of the entropies of activation in the two cases. However, it is evident from the expanded versions of the plot shown here that estimating these entropies (whether graphically or by computer fitting) is equivalent to extrapolating the trends more than 12 times the ranges of the observations. In both cases the data define the centroids with fair accuracy, but, as shown by the shaded envelopes, a very wide range of straight lines could be drawn that would fit the points adequately. (More quantitative accounts of the effects of long extrapolations on parameter estimates may be found in textbooks of statistics, e.g. [6].)

ONE-WAY CATALYSIS

Some reactions, such as the one catalysed by methionine adenosyltransferase, have equilibrium constants that do not favour either the forward or the reverse direction very strongly, and yet have kinetic constants that make them far more effective catalysts for the forward than for the reverse reaction [11]. Such examples are sometimes thought to violate thermodynamic restrictions, but in fact they present no thermodynamic problem, because the equilibrium constant of a reaction only constrains the ratio between the kinetic parameters for the forward and reverse directions, not the individual values, and so large deviations from equality of catalytic constants can be compensated for by equally large deviations in the opposite direction of the corresponding binding constants. This type of consideration was thoroughly discussed by Jencks [12], and more recently by one of us [13].

The equilibrium constant certainly determines the direction in which a reaction will proceed in any particular conditions, and efficient removal of one or more reaction products can ensure that it proceeds in the opposite direction from what naive inspection of the equilibrium constant might suggest. Aspartokinase, for example, catalyses conversion of aspartate and ATP, an anhydride of phosphoric acid, into ADP and phosphoaspartate, a mixed anhydride of phosphoric acid and a carboxylic acid. This is chemically very unfavourable, with an equilibrium constant of 6.4×10^{-4} [14], but the phosphorylation of aspartate still occurs, because the phosphoaspartate is converted so rapidly into aspartic semialdehyde that its concentration is very small (normally too small to measure) in all physiological conditions. In his oral presentation to this symposium Tom Leyh discussed the even more extreme example of the physiological strategy that permits activation of sulphate for use by *Escherichia coli* in spite of an equilibrium constant of the order of 10^{-8}.

In all such cases one needs to take care to avoid interpreting thermodynamic data in an excessively naive way. Although the standard free energy of any reaction unambiguously defines the direction in which the reaction will proceed from the standard state, this will be very misleading if the possibility that the physiological state may be very far from the standard state is not taken into account, and suitable manipulation of one or more product/substrate ratios can ensure that it proceeds in what may appear at first sight to be the wrong direction.

Cornish-Bowden, A. & Cárdenas, M.L.

Nearly Irreversible Reactions in Metabolic Models

Most metabolic reactions have equilibrium constants close enough to unity for it to be imperative to represent them with fully reversible rate equations in metabolic models. However, some proceed virtually irreversibly in the physiological direction of reaction, and it is then important to decide whether they can be safely represented by irreversible rate equations. The reaction catalysed by pyruvate kinase provides the classic example: it is needed for models of glycolysis and has an equilibrium constant of the order of 10^5, and it has divided investigators since the earliest simulations of metabolic systems in the 1960s. Some authors, such as Garfinkel and Hess [15], and, others more recently [16,17], have insisted on the need to use reversible equations throughout, regardless of the magnitudes of the equilibrium constants; others have treated such reactions as irreversible and product-insensitive [18-23].

The question is not trivial, as there are two opposing dangers to be avoided: ignoring reversibility when it must not be ignored leads to an invalid model that makes incorrect predictions; not ignoring it when it is safe to do so leads to a model that is more complicated than it needs to be. The second may appear the lesser of the two dangers, especially in the light of the enormous increases in computing power that have occurred in recent years, but it is a danger nonetheless. Reversible rate equations always contain more parameters than the corresponding irreversible equations, sometimes many more, and in practice this often means that they contain parameters that are not experimentally measured, forcing the modeller to make guesses of dubious validity. When we wished to use a reversible rate equation for pyruvate kinase in a model of glycolysis in *T. brucei* [1,24] we had to face the reality that no experimental data were available for the kinetics of the reverse reaction, forcing us to assume reasonable values that were consistent with the equilibrium constant and with the kinetics of the forward reaction. Even the argument based on consideration of computing power is not entirely convincing, because regardless of the amount of power available the total size of the model accessible to investigation will be decreased by making the individual equations more complicated than they need to be.

Actually, although the examples we have cited represent the dichotomy that existed in the literature between 1964 and 2000-either fully reversible on the one hand, or irreversible and product-insensitive on the other-this is a false dichotomy, because it ignores an intermediate possibility, one that is both real and realistic.

Products affect rates in a fully reversible rate equation in two different and separable ways: they not only cause a negative term to be present in the numerator of the rate expression, allowing the reaction to proceed in reverse under appropriate thermodynamic conditions; they also cause one or more positive terms to appear in the denominator, producing product inhibition even in irreversible conditions.

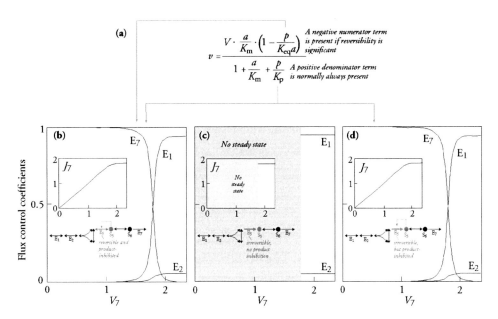

Figure 2. Effects of a product on the rate of an enzyme-catalysed reaction. **(a)** If a product is present in the reaction mixture in sufficient concentrations, it has two different inhibitory effects on the rate of the forward reaction: it leads not only to a negative term in the numerator due to the reversibility of any chemical reaction, but also, and independently, to a positive term in the denominator due to the tendency of almost any product to bind to the substrate-binding site of an enzyme. These properties are conceptually separate, and although the second must normally be present if the first is, it can also be, and often is, important when the first is negligible. The remainder of the Figure illustrates how these considerations act on the behaviour of a metabolic model: **(b)** with both terms present but small in the case of the numerator term; **(c)** with both absent; and **(d)** with only the denominator term present. The part of the model concerned with the action of the product is shown in red. Notice that Panels **(b)** and **(d)** are indistinguishable from one another but very different from Panel **(c)**, so ignoring the denominator term on the mistaken assumption that it can be neglected when the reverse reaction is negligible can lead to gross errors in the behaviour of models. The original article [25] describes the model used for the simulations in detail, and also discusses an additional affect, feedback inhibition by end-product, that we do not consider here.

It follows that although reversibility implies the presence of both kinds of term in almost all cases, irreversibility does not exclude the possibility of product inhibition, and hence a kinetically significant positive denominator term. This point, obvious once pointed out though ignored in the literature for many years, resolves all of the difficulties that we are aware of with pyruvate kinase in metabolic models [25].

In particular, it explains an observation that was initially very puzzling: introducing reversibility of the pyruvate kinase reaction into a model of glycolysis in *T. brucei* [22] caused a far more dramatic redistribution of control than one would have guessed from the small degree of reversibility [24]. Study of more simple models with nearly irreversible reactions in them showed that introducing positive denominator terms into a rate equation could lead to significant changes in behaviour, whereas introducing negative numerator terms that would be extremely small in any reasonable conditions never produced significant effects [25]. There is, of course, nothing surprising in this result; all that is surprising is that it was ignored for so long. As a general conclusion, it appears clear that one can safely ignore reversibility (negative numerator terms) when modelling nearly irreversible reactions, but it will not be safe to ignore product inhibition (positive denominator terms) unless there is very good experimental evidence that product inhibition is truly negligible over the whole range of conditions to be simulated.

An important point to consider is the possible difference between isoenzymes, which despite catalysing the same reaction, and thus having the same thermodynamic constraints, may vary considerably in the importance of reversibility and product sensitivity in the reaction, because of differences in kinetic parameters. This is well illustrated by the mammalian hexokinases [26]. Product inhibition of the liver isoenzyme, hexokinase D, is always negligible, as the inhibition constant for the product glucose 6-phosphate is far above any conceivable physiological concentrations, but inhibition of the other isoenzymes by glucose 6-phosphate is by no means negligible. Furthermore, although the equilibrium constant of the reaction, phosphorylation of glucose by MgATP, is not high enough to make the negative term in the numerator necessarily negligible, this term can certainly be neglected for hexokinase D because the concentration of glucose 6-phosphate needed to half-saturate the enzyme is very high.

In all of this discussion we are, of course, concerned with reactions that are essentially irreversible in the physiological direction. A reaction like that of aspartokinase, considered earlier, must of course be treated as reversible unless the rest of the model ensures that one or more product concentrations are truly negligible, because although they have large equilibrium constants these do not favour the direction in which the reaction is proceeding.

DISCUSSION

Thermodynamic ideas remain the basis of much of physical chemistry and of biochemistry. Avoiding the simple errors that we discussed at the beginning of this article is not difficult, but it does require more than a superficial understanding of what is being measured and what the numbers mean. The growing interest in systems biology implies a growing need to incorporate information about the kinetics and thermodynamics of enzyme reactions into large-scale models of cells. This certainly involves avoiding simple errors, of course, but it also demands intelligent decisions to be made to deal with the lack of much of the experimental information that one would ideally want to incorporate in a model. At present kinetic information is so sparse for some organisms that one is forced to develop methods that allow predictions based solely on the stoichiometric structure of a system [27,28], but this will certainly improve in the future if the enzymes concerned are studied directly. Even systems, for which a large amount of experimental kinetic information exists, such as the glycolytic pathway in *T. brucei* [22], have some gaps that need to be filled with reasonable guesses. In such cases knowing that nearly all reactions (including those that are virtually irreversible) are sensitive to their products can allow models that avoid being excessively simple, and hence capable of making invalid predictions, while also avoiding being excessively complicated, requiring values to be unnecessarily guessed and equations to be unnecessarily complicated. A simple rule in such cases is that it is safe to omit the term in the numerator that represents a reverse reaction that occurs to a negligible extent, but it is not safe to omit the terms in the denominator that take account of product inhibition. If any such simple rule is used, however, one must take account of the very different conditions that exist in different tissues and in different metabolic states, as well as the different kinetic parameters of isoenzymes that catalyse the same reaction.

REFERENCES

[1] Cornish-Bowden, A., Cárdenas, M.L. (2003) *Molecular Informatics: Confronting Complexity* (Hicks, M.G., Kettner, C., Eds), pp. 95-107. Beilstein, Frankfurt.

[2] Cooper, L.C. (1969) *An Introduction to the Meaning and Structure of Physics*. Harper & Row, Tokyo.

[3] Laidler, K.J. (1993). *The World of Physical Chemistry*, pp. 83-130. Oxford University Press, Oxford.

[4]   Velázquez-Campoy, A., Leavitt, S.A., Freire E. (2004) Characterization of protein-protein interactions by isothermal titration calorimetry. *Methods Mol. Biol.* **261**:35-54.

[5]   Sievers, A., Beringer, M., Rodnina, M.V., Wolfenden, R. (2004) The ribosome as an entropy trap. *Proc. Natl. Acad. Sci. USA* **101**:7897-7901.

[6]   Colquhoun, D. (1971) *Lectures on Biostatistics*, pp. 228-234. Clarendon Press, Oxford.

[7]   Johnson, I.A., Goldspink, G. (1975) Thermodynamic activation parameters of fish myofibrillar ATPase enzyme and evolutionary adaptations to temperature. *Nature* **257**:620-622.

[8]   Gutfreund, H. (1995) *Kinetics for the Life Sciences*, pp. 246-248. Cambridge University Press, Cambridge.

[9]   Exner, O. (1964) On the enthalpy-entropy relationship. *Coll. Czech. Chem. Comm.* **26**:1094-1113.

[10]   Cornish-Bowden, A. (2002) Enthalpy-entropy compensation: a phantom phenomenon. *J. Biosci.* **27**:121-126.

[11]   Mudd, S.H., Mann, J.D. (1963) Activation of methionine for transmethylation. VII. Some energetic and kinetic aspects of the reaction catalyzed by the methionine-activating enzyme of bakers' yeast. *J. Biol. Chem.* **238**:2164-2170.

[12]   Jencks, W.D. (1975) Binding energy, specificity, and enzymic catalysis: the Circe effect. *Adv. Enzymol.* **43**:219-410.

[13]   Cornish-Bowden, A. (2004) Fundamentals of Enzyme Kinetics, 3rd Edn, pp. 53-55. Portland Press, London.

[14]   Chassagnole, C., Rais, B., Quentin, E., Fell, D.A., Mazat, J.-P. (2001) An integrated study of threonine-pathway enzyme kinetics in *Escherichia coli*. *Biochem. J.* **356**:415-423.

[15]   Garfinkel, D., Hess, B. (1964) Metabolic control mechanisms VII. A detailed computer model of the glycolytic pathway in ascites cells. *J. Biol. Chem.* **239**:971-983.

[16]   Schuster, R., Holzhütter, H.-G. (1995) Use of mathematical models for predicting the metabolic effect of large-scale enzyme activity alterations. Application to enzyme deficiencies of red blood cells. *Eur. J. Biochem.* **229**:403-418.

[17]   Mulquiney, P.J., Kuchel, P.W. (1999) Model of 2,3-bisphosphoglycerate metabolism in the human erythrocyte based on detailed enzyme kinetic equations: equations and parameter treatment. *Biochem. J.* **342**:581-596.

[18]   Heinrich, R., Rapoport, T.A. (1973) Linear theory of enzymatic chains: its application for the analysis of the crossover theorem and of the glycolysis of the human erythrocyte. *Acta Biol. Med. Germ.* **31**:479-494.

[19]   Heinrich, R., Rapoport, T.A. (1974) A linear steady-state theory of enzymatic chains: general properties, control and effector strength. *Eur. J. Biochem.* **42**:89-95.

[20]   Joshi, A., Palsson, B.O. (1990) Metabolic dynamics in the human red cell. Part III-Metabolic reaction rates. *J. Theor. Biol.* **142**:41-68.

[21]   Heinrich, R., Schuster, S. (1996) *The Regulation of Cellular Systems*, pp. 170-188. Chapman & Hall, New York.

[22]   Bakker, B.M., Michels, P.A.M., Opperdoes, F.R., Westerhoff, H.V. (1997) Glycolysis in bloodstream from *Trypanosoma brucei* can be understood in terms of the kinetics of the glycolytic enzymes. *J. Biol. Chem.* **272**:3207-3215.

[23]   Bakker, B.M., Michels, P.A.M., Walsh, M.C., Opperdoes, F.R., Westerhoff, H.V. (2000) In: *Technological and Medical Implications of Metabolic Control Analysis*, (Cornish-Bowden, A., Cárdenas, M. L., Eds), pp. 157-164. Kluwer, Dordrecht.

[24]   Eisenthal, R., Cornish-Bowden, A. (1998) Prospects for antiparasitic drugs: the case of *Trypanosoma brucei*, the causative agent of African sleeping sickness. *J. Biol. Chem.* **273**:5500-5505.

[25]   Cornish-Bowden, A., Cárdenas, M.L. (2001) Information transfer in metabolic pathways: effects of irreversible steps in computer models. *Eur. J. Biochem.* **268**:6616-6624.

[26]   Cárdenas, M.L. (1995) *"Glucokinase": its Regulation and Role in Liver Metabolism*, pp. 21-40. R. G. Landes, Austin, Texas.

[27]   Stelling, J., Klamt, S., Bettenbrock, K., Schuster, S., Gilles, E.D. (2002) Metabolic network structure determines key aspects of functionality and regulation. *Nature* **420**:190-193.

[28]   Cornish-Bowden, A., Cárdenas, M.L. (2002) Metabolic balance sheets. *Nature* **420**:129-130.

The Chemical Theatre of Biological Systems, May 24th - 28th, 2004, Bozen, Italy

Navigation in Chemical Space Based on Correlation-Vector Representation of Molecules

Gisbert Schneider*, Steffen Renner and Uli Fechner

Johann Wolfgang Goethe-Universität, Beilstein Endowed Chair for Cheminformatics,
Institute of Organic Chemistry and Chemical Biology,
Marie-Curie-Str. 11, D-60439 Frankfurt am Main, Germany

E-Mail: *gisbert.schneider@modlab.de

Received: 20th September 2004 / Published: 22nd July 2005

Abstract

Correlation-vector representation (CVR) of molecular structure and properties results in an alignment-free descriptor. This facilitates rapid virtual screening of large virtual compound libraries and chemical databases. The approach has a tradition in chemoinformatics and has already led to the identification of several new lead structures. Its foremost application is ligand-based design of activity-enriched, focused compound libraries. Before applying CVR it is essential to consider appropriate descriptor scaling, select a suitable similarity metric and choose meaningful reference molecules. It was demonstrated that there exists no cure-all recipe for this task. Both three-dimensional and two-dimensional CVR and different similarity metrics complement each other yielding an improved hit rate of the combined approach.

Introduction

The quest for novel drugs might be considered as a "journey through chemical space", and in order to be successful, we need a navigation system - unless we are satisfied with a random walk. Navigation can be defined as "The process of determining and maintaining a course or trajectory to a goal location" [1].

This means that we need:

- a coordinate system that defines the search space. A convenient way to do this is to employ molecular descriptors, which can be used to generate molecular encoding schemes reaching from general properties (e.g. lipophilicity, molecular weight, total charge, volume in solution, etc.) to very specific structural and pharmacophoric attributes (e.g. multi-point pharmacophores, field-based descriptors).
- a target location. Sets of reference molecules, or "seed structures", exhibiting a desired biological activity (ligand-based approach), or a model of the binding pocket of the macromolecular target receptor (structure-based approach) serve this purpose. The aim is to find "activity islands" in chemical space that are populated by molecules that are similar to the reference compounds (i.e., they are found in a neighbourhood of the reference structures), or can be predicted to bind to the target receptor.
- a method of moving in search space. Typically, sampling methods ("cherry picking") are employed for focused library design, or *de novo* design approaches to identify novel candidate molecules.
- a map. The map separates the search space into regions of high and low quality. This provides the basis for a directed movement toward the target location. Quantitative Structure-Activity (QSAR) models, property prediction methods, and scoring functions - just to name some possibilities - can be used. It should be stressed that each map has a certain resolution and meaning, and depending on the definition of search space and the aim of our journey, we will have to use different maps and navigation systems. Systematic navigation with appropriate maps can not only help us find potential new lead structures but also tell us something about the expected activity profile of a candidate molecule, which is desirable to make an informed selection and prioritization of candidate leads with a reduced attrition liability.

Most critical are the choice of the search space coordinates, and the map resolution. Molecules must be represented in a suitable fashion for reliable prediction of molecular properties [2]. In other words, the appropriate level of abstraction must be defined to perform rational virtual screening.

"Filtering" tools can then be constructed using a simplistic model relating the descriptors to some kind of bioactivity or molecular property. However, the selection of appropriate descriptors for a given task is not trivial and careful statistical analysis is required. Besides an appropriate representation of the molecules under investigation, any useful feature extraction system must be structured in such a way that meaningful analysis and pattern recognition is possible. Technical systems for information processing are intuitively considered as mimicking some aspects of human capabilities in the fields of perception and cognition. Despite great achievements in artificial intelligence research during the past decades and an increasing application of machine learning methods in virtual screening such as Artificial Neural Networks (ANN) and Support Vector Machines (SVM) [3], we are still far from understanding complex biological information processing systems in detail. This means that a feature extraction or "navigation" task that appears simple to a human expert can be extremely hard or even impossible to solve for a technical system, e.g. particular virtual screening software. As we have learned from many years of "artificial intelligence" research, it is extremely difficult (if not impossible) to develop virtual screening algorithms mimicking the medicinal chemists' intuition. Furthermore, there is no common "gut feeling" because different medicinal chemists have different educational backgrounds, skills and experience. Despite such limitations there is, however, substantial evidence that it is possible to support drug discovery in various ways with the help of computer-assisted library design and selection strategies. There are two specific properties of computer-based approaches which make them very attractive for exploratory navigation in chemical space, namely their speed of execution which can be significantly faster than in vitro experiments, and the ability to access virtual compound libraries. By this means "unexplored" regions of chemical space can be entered and analysed.

Due to its ease of implementation, chemical similarity searching has a long tradition in this area, and many different similarity metrics have been proposed to analyse rapidly very large virtual libraries [4-7]. Again, similarity searching can only be successful if molecules are represented by a suitable description of the chemical space. The definition of "important" attributes heavily depends on the query structure and therefore on its associated binding partner. Descriptors of chemical space can be categorized, e.g., according to their data representation and according to the dimensionality of molecular attributes (1D, 2D or 3D) they describe. Binary fingerprints are a typical data representation for similarity searching [8].

They describe the presence or absence of a feature, e.g. a substructure, or a certain pharmacophore, in a linear bit string format. Fingerprints vary in length from 57 bits for mini-fingerprints (a collection of 1D and 2D molecular descriptions [9]) up to millions of bits of 4D-pharmacophore fingerprints (all combinations of four-point pharmacophores) [10]. For an extensive review of issues related to conformer generation in the process of property calculations, see elsewhere [11].

Pharmacophore models seem to be specifically suited for "scaffold hopping" and library design [12]. If we wish to pick members of a compound library from a very large virtual chemistry space, the calculation of 3D-conformers and subsequent structural or potential pharmacophore point-based alignment of molecules can be a time-limiting factor. Therefore, alignment-free models have a value particularly during the early phases of library design [13-15]. To demonstrate the idea, one representative of these methods shall be discussed in more detail: correlation-vector representations (CVR). The correlation vector approach was introduced to the field of cheminformatics by Broto and Moreau over two decades ago [16], and brought to wider attention through studies by Gasteiger and coworkers [17-19]. The basic idea of CVR is to map molecular features, e.g. pharmacophore points or properties, to a numerical vector of fixed length which is similar to fingerprint generation. As a consequence, such a vector of a given dimension encodes each molecule, and pair-wise comparison of vectors (e.g., by similarity calculation) can be executed very quickly without having to explicitly align the molecular structures. CVR belongs to the class of alignment-free descriptors. Our group and others have reported several CVR applications to similarity searching previously, exploiting the possibility for very fast virtual screening of large compound collections or in *de novo* design [20]. Here we give an overview of our contributions to this field.

SIMILARITY SEARCHING USING CORRELATION-VECTOR REPRESENTATIONS

Chemical similarity searching can serve as a guide through chemical space with the goal of identifying novel molecules that reveal similar biological activity as a query structure. It is often employed in early-phase virtual screening for the selection of activity-enriched subsets [21]. Ligand-based similarity measures facilitate similarity searching in the absence of receptor structure information and are frequently and successfully used for this purpose [22-24].

Basically, these methods rely on:

- one or more representative reference structures ('query structures'),

- molecular descriptors that capture biologically relevant attributes,

- a suitable similarity metric.

The foundation of chemical similarity searching is the pair-wise compound comparison between the query structure and the compounds of a screening library. This reflects the underlying supposition that structurally similar molecules have similar biological activities [25]. The result of a similarity search is a similarity-ranked list. High-ranking compounds in such a list are assumed to be more similar to a query structure than low-ranking compounds. The similarity metric is responsible for considering molecules as different, which do not share important attributes. The definition of 'important' attributes crucially depends on the query structure and thus on the respective biological target.

A similarity search can be performed either retrospective (retrospective screening) or prospective (prospective screening). Retrospective screening is carried out with a set of molecules that are active against a certain biological target (query structures) and a screening library that is compiled of compounds that are inactive against the same target. (Most datasets suffer from *assumptive inactivity*: because of the non-inexistence of measured data it is often unknown whether the inactive compounds are really inactive against the particular biological target). For each of the n known actives of a given dataset a pair-wise similarity search is performed. Hence, there are n similarity searches where each of the known actives in turn, is the query structure. The pair-wise similarity of the respective query structure is calculated against all the remaining known actives (n-1) and all compounds of the screening library (inactive compounds). This procedure yields n similarity ranked lists that are ultimately fused into a final similarity ranked list that incorporates the rankings of the n individual lists. There exist different ideas of how to combine the individual lists [26]. Retrospective screening starts with the a priori knowledge of which compounds are active (query structures) and which are not (screening library). This knowledge is then consulted to assess the quality of the final similarity ranked list. Thus, retrospective screening does not aim to come up with novel molecular structures. Its aim is rather to evaluate the quality of the applied parameters, i.e., the molecular descriptor and the similarity metric. It is noteworthy that the dataset (query structures and screening library) also has a prominent influence on the screening results. This limits the explanatory power of comparisons of virtual screenings with different datasets.

The descriptors and similarity metrics that were tested in a retrospective screening can then be applied to prospective screening. Here, one starts with one or more known active molecules (query structures) as well. However, the screening library is distinct from the one employed with retrospective screening. The intended purpose of prospective screening is to identify molecules within the screening library that exhibit activity against the same biological target as the query structures. Therefore, the screening library should contain compounds that are likely to exhibit that similarity.

The quality of a retrospective screening is often quantified by the enrichment factor (*ef*):

$$ef = \left(\frac{S_{act}}{S_{all}}\right) \Big/ \left(\frac{D_{act}}{D_{all}}\right) \tag{1}$$

The enrichment factor is calculated for defined fractions (subsets) of the similarity ranked list, e.g. the first percentage, the first two percentages etc. D_{all} is the total number of compounds in the dataset (query structures and screening library), and S_{all} is the number of molecules in the subset. D_{act} is the number of known active molecules (query structures) in the dataset, and S_{act} is the number of actives found in the subset. The plotting of S_{all}/D_{all} on the abscissa and S_{act}/D_{act} on the ordinate leads to the visualization of the enrichment factor, i.e. the enrichment curve. A method that is superior to a random selection of compounds returns an *ef* > 1 and an enrichment curve above the diagonal line. It should be emphasized that the enrichment factor has an upper limit that is contingent on the fraction of active compounds in the dataset. Again, this stresses the fact that a comparison of several retrospective-screening runs is only meaningful if they are carried out with the same dataset.

In a recent study we investigated the influence of individual parameters on ligand-based virtual screening [27]. This influence was examined on three levels: On the level of the dataset, the descriptor, and the similarity metric. For this purpose, we employed twelve different datasets, three different molecular representations and three different similarity metrics (the Manhattan Distance, the Euclidian Distance and the Tanimoto Coefficient). Special focus was on the evaluation of CVRs of molecular features [16]. The basic idea of CVR is to map molecular features, e.g. pharmacophore points, to a numerical vector of fixed length. Since CVRs have the convenience of being alignment-free descriptors, pair-wise comparisons of vectors can be executed very quickly.

The twelve datasets were compiled from the COBRA database [28] and consisted of a set of active compounds (query structures) and the respective remainder of the COBRA database as 'inactive compounds' (screening library). All molecules of one set of active compounds bind to the same interaction partner, but the definition of these interaction partners differs between the individual sets. These distinct levels of specificity range from receptor classes that comprise a rather diverse set of molecules (e.g. GPCR) to particular receptor subtypes (e.g. COX2). The twelve sets of active compounds contain ligands that bind to angiotensin converting enzyme (ACE, 44 compounds), cyclooxygenase 2 (COX2, 93 compounds), corticotropin releasing factor (CRF antagonists, 63 compounds), dipeptidyl-peptidase IV (DPP, 25 compounds), G-protein coupled receptors (GPCR, 1642 compounds), human immunodeficiency virus protease (HIVP, 58 compounds), nuclear receptors (NUC, 211 compounds), matrix metalloproteinase (MMP, 77 compounds), neurokinin receptors (NK, 188 compounds), peroxisome proliferator-activated receptor (PPAR, 35 compounds), beta-amyloid converting enzyme (BACE, 44 compounds) and thrombin (THR, 188 compounds).

All datasets were encoded by three different descriptors: CATS2D [29], CATS3D and CHARGE3D. The two CATS descriptors belong to the category of atom-pair descriptors. The centres of the atom-pairs are not characterized by their chemical element type but by their membership to a potential pharmacophore point (PPP) group. CATS2D considers five PPP groups: hydrogen-bond donor, hydrogen-bond acceptor, positively charged or ionized, negatively charged or ionized, and lipophilic. The PPPs of CATS3D were associated with the PATTY-Type function of MOE [30] that closely follows the assignment scheme of the PATTY publication by Bush and Sheridan [31]. This assignment scheme comprises seven PPPs: cationic, anionic, polar, acceptor, donor, hydrophobic and other. All possible pairs of PPPs were then interrelated with the distance between the corresponding atoms. Whereas the CATS2D descriptor regards topological distances measured in bond lengths, the CATS3D descriptor incorporates the spatial Euclidian distance of atom pairs. Both CATS descriptors were scaled to diminish the influence of the molecular size. CHARGE3D is based on the correlation vector approach of Gasteiger and co-workers [17]. Calculated partial atom charges are assigned to all atoms. To yield a single charge value for each possible atom pair of a molecule the charge values of the two respective atoms are multiplied.

Finally, the single charge value of each atom pair is associated with the spatial Euclidian distance between the two atoms to obtain a high-dimensional correlation vector. In case of CATS3D and CHARGE3D descriptor a single conformation for each molecule of the dataset was calculated with the program CORINA [32].

Enrichment factors between 2 (GPCR dataset) and 26 (CRF antagonists) were acquired for the first percentage of the datasets (a complete list of the enrichment factors can be found in the original publication). Aside from the GPCR dataset considerable enrichment factors were achieved with all three descriptors. In most cases the influence of the similarity metric was marginal, albeit there were a few combinations of datasets and descriptors that showed a significant discrepancy in terms of the enrichment factor with different similarity metrics. Figure 1 illustrates the enrichment curves for the COX2, the HIV, and the MMP subsets of the COBRA database with the Manhattan Distance as a distance metric. Many active compounds received top ranks with CATS2D and CATS3D for all three datasets depicted in Fig. 1. Whilst COX2 ligands were successfully enriched with the CHARGE3D, after approximately 20% of the screened dataset the enrichment curves for the HIVP and MMP ligand datasets drop below the curve that represents a random distribution of active compounds in the dataset. Nevertheless, HIVP and MMP ligands were enriched by CHARGE3D in the first percentiles of the datasets. Figure 1 clearly demonstrates that none of the descriptors is superior for all three datasets, but there is a preferred one for a given dataset. The suitability of the descriptors depends on the underlying dataset, i.e. the binding patterns of a specific ligand-receptor pair. Distinct performances of the descriptors were expected, as the CATS2D encodes topological information of PPPs, the CHARGE3D three-dimensional information of partial atom charges and the CATS3D spatial information of PPPs.

The separation of active and inactive compounds performed variably contingent on the dataset. Irrespective of the descriptor and similarity metric the approximate classification accuracy seems to be determined by the dataset. Some target classes yielded better enrichment factors than others. We deduced two possible reasons for this behaviour. First, the three descriptors may cover the essential binding pattern of particular datasets to a different extent. Second, the individual datasets are defined at different levels of specificity. The latter hypothesis is substantiated by the fact that GPCR ligands could not be considerably enriched but significant enrichments were achieved for stricter defined subsets of GPCRs. For example, enrichment factors for CRF range from 9 to 26 in the first percentile.

Hence, the separation of actives and inactives may even be possible in difficult cases provided the definition of these two classes is specific enough. Again, this emphasizes that the dataset with its inherent properties has a major influence on the outcome of a virtual screening campaign.

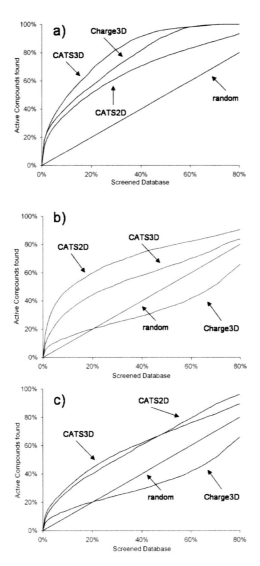

Figure 1. Enrichment curves with the **(a)** COX2, **(b)** HIV protease and **(c)** MMP subset of the COBRA database as active molecules based on the Manhattan distance. The curve indicated with 'random' results from a random distribution of the actives (query structures) among the inactive molecules (screening library).

Whereas the enrichment factor discriminates only between active and inactive compounds, we then investigated which active compounds were retrieved by the three descriptors among the top-ranking ones. Figure 2 depicts this for the first five percentiles of the COX2, MMP and HIVP dataset by means of Euler-Venn diagrams. It is noteworthy that although the enrichment factors with different descriptors were approximately the same, the active compounds among the top-ranking ones varied. Figure 2 shows that a large number of compounds were exclusively retrieved with a single descriptor and that the intersection sizes of all three descriptors were rather small. These two observations sustained the hypothesis that each descriptor covers a certain, and to a varying extent different, aspect of the ligand-receptor binding pattern. Moreover, the information contents of the three descriptors complement each other. The extent of completion can be measured by calculation of the "cumulative percentages" for a given dataset: for a given dataset the active compounds among the first 5% of the similarity-ranked lists of the three descriptors are extracted to obtain three sets of active compounds. The sets are then united and the number of elements of the united set is related to the total number of active compounds of the particular dataset. Cumulative percentages facilitated the retrieval of additional 7% to 52% of active compounds compared to the exclusive employment of the CATS2D. Thus, it may be appropriate to unite the information encoded by different descriptors if a similarity search is performed to cover more facets of the ligand-receptor binding pattern under investigation.

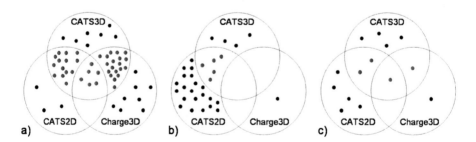

Figure 2. Elements of the Euler-Venn diagrams correspond to compounds that are retrieved among the first 5% of the similarity ranked list that results from retrospective screening with the **(a)** COX2, **(b)** HIV protease and **(c)** MMP subsets of the COBRA database as actives. The Manhattan distance was employed as a distance measure. Membership indicates that the respective compound was retrieved by retrospective screening with the corresponding descriptor.

Another study focused on the comparison of seven similarity metrics for ligand-based similarity searching [33]. The same twelve datasets compiled of the COBRA database as in the aforementioned experiments were employed.

All datasets were encoded with the CATS2D descriptor and retrospective screenings were carried out with seven similarity metrics: Manhattan Distance, Euclidian Distance, Tanimoto Coefficient, Soergel Distance, Dice Coefficient, Cosine Coefficient, and Spherical Distance. Again, apart from the GPCR dataset considerable enrichments were achieved. Enrichment factors for the same datasets but different similarity metrics differed only slightly. For almost all datasets, the Manhattan and the Soergel Distance yielded the overall highest enrichment factors. One might conclude that if only a single distance metric is applied the Manhattan Distance should be preferred due to its computational simplicity and altogether above-average performance.

The study also addressed the question to which extent the active compounds among the top-ranking ones were identical if different similarity metrics were applied. Each of the twelve datasets yielded seven similarity ranked lists obtained with the seven similarity metrics. For each dataset the cumulative percentages were calculated for the first 5% of these lists. This procedure led to the retrieval of significantly more hits than found by any single similarity metric. The increase of the cumulative percentages for all seven metrics compared to the employment of only the Manhattan Distance ranged from additional 5% (COX2) to 28% (NUC and MMP) with an average of 19% over all twelve datasets.

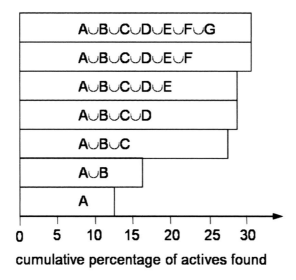

Figure 3. Bars indicate the "cumulative percentage" of active compounds (ligands of nuclear receptors, NUC, from the COBRA data collection) found among the top-ranking 5% of the similarity-ranked list that results from a retrospective screening. **A)** Manhattan distance, **B)** Euclidian distance, **C)** Tanimoto coefficient, **D)** Soergel distance, **E)** Dice coefficient, **F)** Cosine coefficient, **G)** Spherical distance.

Figure 3 illustrates this gradual rise for the NUC dataset. This study suggests that not only different descriptors complement each other but also different similarity metrics. Therefore, it might be advantageous to employ several molecular descriptors and similarity metrics in parallel and benefit from a unification of the various definitions of "chemical similarity".

CODING RECEPTOR INFORMATION INTO CORRELATION-VECTOR REPRESENTATION

Different types of reference points can guide navigation through chemical space. So far, we have pinpointed approaches that employed ligand information to reach our goal location. But if the receptor structure of a biological target is available it can also direct our search for novel ligands. Classical approaches in this field are molecular docking tools such as FlexX or GOLD [34,35]. One drawback with these methods is their prolonged calculation time compared to topological ligand-based methods, due to conformer calculation and spatial feature alignment. Our research group developed a method that combines the speed advantages of ligand-based similarity searching with the ability to exploit binding pocket information (A. P. Schüller, U. Fechner, S. Renner, L. Franke, L. Weber, G. Schneider, unpublished). This is accomplished by introducing the "virtual ligand" (VL). The VL represents a super-structure of potentially all conceivable ligands of a binding pocket. A three-dimensional binding pocket model serves as a starting point to define three types of receptor-based PPPs (hydrogen-bond acceptor, hydrogen-bond donor and lipophilic). These receptor-based PPPs are then used to create potential interaction sites in the cavity of the binding pocket to form a VL. The interaction types of the interaction site are complementary compared to those of the receptor-based PPPs (hydrogen-bond donor is converted to hydrogen-bond acceptor and vice versa, the lipophilic type is unchanged). Encoding the VL as a correlation vector descriptor circumvents computationally expensive 3D alignment of molecular features of the VL and candidate structures. The three interaction types and 20 equidistant bins yielded a 120-dimensional descriptor. For such a calculated VL three parameters were optimized with ten-fold cross-validation prior to virtual screening (scaling of the descriptor, a weighting scheme and a distance metric). As the parameter optimization was performed on a set of reference compounds the method is not solely receptor-based but may be regarded as a hybrid-approach that includes both receptor and ligand information. Up to 50 conformations were calculated for each compound in the screening library with the Molecular Operating Environment (MOE) conformer generator [30]. The MOE atom typing (PATTY types) was used to assign PPPs to the structures in the screening library [31]. Subsequent application of the same correlation function as used for the calculation of the VL, generated a 120-dimensional descriptor.

Several retrospective and one prospective screening study were performed to provide a proof-of-concept for this approach. All screenings included a randomized VL as a negative control to determine the meaningfulness of the receptor-derived VLs. Co-crystal structures of the PDB served as a starting point for the calculation of the VL. The first series of experiments was considered as retrospective screening because the reference compounds and the screening library were part of the same dataset. Two retrospective screenings with the Factor Xa and COX-2 subset of the COBRA database as actives and the remainder of the database as inactives resulted in significant enrichments. Comparable results were achieved with the Factor Xa inhibitors of a compilation of 15,840 products of the three-component Ugi-reaction, that were tested for inhibition of five serine proteases {courtesy of Dr L. Weber, Morphochem AG, Munich, Germany, unpublished}. Moreover, the receptor-based VL perspicuously outperformed the randomized VL in all three screenings. A final prospective screening was carried out with the Factor Xa dataset of the COBRA database for parameter optimization and the Factor Xa ligands in the Ugi dataset for similarity searching. Here, measured activity values of the Ugi dataset were only employed to assess the quality of the similarity-ranked lists. Among the 50 top-ranking compounds four molecules exhibited a K_i below 10 µM.

The outcome of these experiments demonstrates that the VL approach is suited to retrieve active chemotypes from a library of reference compounds by means of chemical similarity searching. The encompassment of information from binding pockets and their known ligands, bridges the gap between structure-based and ligand-based virtual screening methods.

FUZZY PHARMACOPHORES: THE SQUID APPROACH

Using information from descriptors representing single active molecules is one possible way to define reference points in chemical space. An alternative approach is to use pharmacophore models which comprise information from multiple active molecules within one model. This approach can be characterized as an "ideal ligand" approach [36]. In comparison to many machine learning-, clustering- or consensus scoring-techniques a pharmacophore model is typically calculated from an explicit alignment of molecules and not from the feature space of multiple single molecule descriptors.

A pharmacophore model represents the spatial configuration of generalized interaction sites, which are essential for biological activity [36,37].

New molecules, which comprise these features, are assumed to be active. Usually these interaction sites represent the most conserved features of a set of known active molecules, which are not present in inactive molecules.

Based on the alignment of active molecules, tolerances for the features are usually estimated to compensate for ligand and receptor flexibility. SQUID fuzzy pharmacophore models extend this approach by approximation of an alignment of known active ligands by a set of spatially distributed Gaussian probability densities for the presence of pharmacophoric features [38]. Features that are present in many of the reference molecules result in a high probability and features which are rarely present in the underlying molecules result in a low probability. Tolerances of the features, which are considered by this approach, might be better represented by Gaussian densities than by rigid spheres. For the resulting fuzzy pharmacophore models different degrees of fuzziness can be defined, e.g. the model can be very generalized or more restricted to the underlying distribution of atoms from the alignment.

For virtual screening the three-dimensional spatial distribution of Gaussian densities is transformed into a two-point correlation vector representation which describes the same probability density for the presence of atom pairs, comprising defined pharmacophoric features. This representation is independent from translation and rotation which makes rapid database screening possible without the necessity to explicitly align the molecules, which can be a limiting step for the screening of large databases. This renders the fuzzy pharmacophore CVR useful for ranking 3D pharmacophore-based CVR representations of molecules, namely CATS3D descriptors of molecules [27]. Consequently SQUID can be characterized as a hybrid approach between conventional pharmacophore searching, similarity searching and fuzzy modelling.

An overview over the calculation of the pharmacophore model and the CVR is given in Fig. 4. The starting point is an alignment of known active ligands. Each atom of the molecules is assigned to a general pharmacophoric atom type like hydrogen-bond donor, hydrogen-bond acceptor or hydrophobic, which results in a field of features.

Maxima in the local feature densities (LFD) are used as cluster seeds to cluster the features into potential pharmacophore points (PPPs) for a more general representation of the underlying alignment. The degree of abstraction, generality or fuzziness of the resulting model can be defined by the cluster radius, a variable which affects the calculation of the LFD and the radius within which maxima are determined.

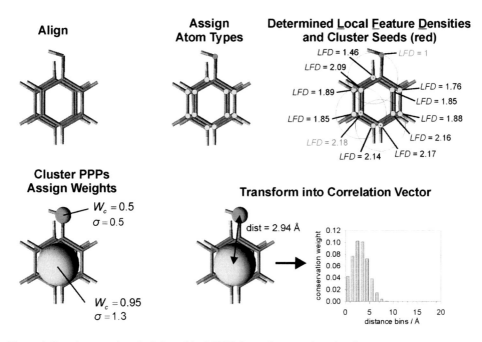

Figure 4. Overview over the calculation of the SQUID fuzzy pharmacophore descriptor.

The radius or standard deviation of the PPPs is dependent on the distribution of the atoms, which are clustered into each particular PPP. Conservation weights (Wc) of the PPPs quantify the conservation of the pharmacophoric features of each of the PPPs in the underlying alignment. As the last step, the three-dimensional representation of the fuzzy pharmacophore model is transformed into a correlation vector for virtual screening. The distance dependent probability for all pairs of PPPs was calculated and subdivided by distance bins and pharmacophoric feature types, resulting in the SQUID CVR.

Figure 5 shows an alignment of the COX-2 inhibitors M5, SC-558 and Rofecoxib, which was adapted from Palomer and coworkers [39]. According to Palomer et al. essential interactions for specific COX-2 inhibition, are mediated by the aromatic rings A and B and the sulfonyl group.

A set of pharmacophore model representations were calculated with different cluster radii resulting in models with different degrees of fuzziness. The model with 1 Å cluster radius resulted in the most detailed representation of the underlying alignment, accordingly with the lowest abstraction from the scaffolds of the molecules in the alignment.

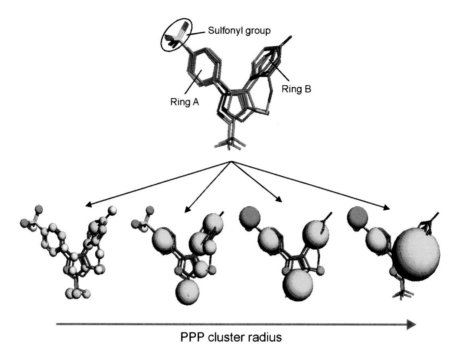

Figure 5. Alignment of three COX-2 inhibitors and SQUID fuzzy pharmacophore models with different resolutions. From left to right, cluster radii of 1.0 Å, 1.5 Å, 2.5 Å and 3.5 Å were used for the model calculation.

Using larger cluster radii results in pharmacophore models with higher degrees of generalization until, like the model resulting from 3.5 Å, the underlying alignment is only marginally visible. For virtual screening it has to be tested for each target and each set of molecules in an alignment from which the degree of fuzziness results in molecules which are most likely to be active, or possess some desired characteristics. Retrospective screening for known active molecules with models with different resolutions provides one possible rationale for that goal.

In Figure 6 the CVR of the best-found COX-2 fuzzy pharmacophore model [38] with a cluster radius of 1.4 Å is shown in comparison to the scaled CATS3D vectors of the underlying molecules from the alignment.

As one can see the fuzzy and thus "generalizing" representation of the underlying molecules from the alignment is retained in the CVR. It becomes clear that the SQUID CVR and the CATS3D CVRs differ significantly in the meaning of their content.

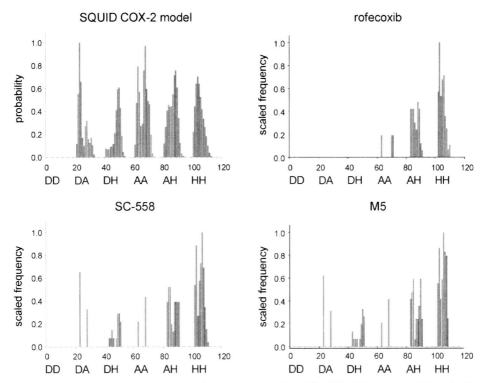

Figure 6. Correlation vector representations of the SQUID COX-2 model and CATS3D vectors of the molecules from the alignment used for the calculation of the pharmacophore model.

The SQUID CVR describes a broad range of descriptor areas which are favourable for the desired biological activity, while the CATS3D descriptor contains only a smaller subset of the actual occurrences of atom-pairs in a specific ligand. Consequently, commonly used similarity metrics like the Euclidean distance or the Tanimoto index, which are based on the assumption that both descriptors, which are to be compared, represent objects in the same way, cannot be used to assess the activity of the molecules under consideration. To overcome this problem a SQUID similarity score was developed (Equation 2):

$$S(a,b) = \frac{\sum\limits_{i=1}^{n}(a_i * b_i)}{1 + \sum\limits_{i=1}^{n}((1-a_i)*b_i)} \qquad (2)$$

where a_i is the value of the i-th element of the SQUID CVR, b_i is the value of the i-th element of a molecule CVR and n is the total number of dimensions. The value a_i may be considered as the idealized probability of the presence of features in b_i.

This results in high scores for molecules with many features in regions of the query descriptor which have a high probability. To penalize the presence of such atom pairs in regions with a low probability, the denominator weights the presence of atom pairs with the inverted probabilities of the descriptor of the pharmacophore model (a value of 1 was added to the denominator to avoid division by zero and high scores resulting from a very low value in the denominator of the term). Accordingly the SQUID scores for the CVRs from the COX-2 inhibitors in Fig. 6 decrease in the order of M5 > SC-558 > Rofecoxib.

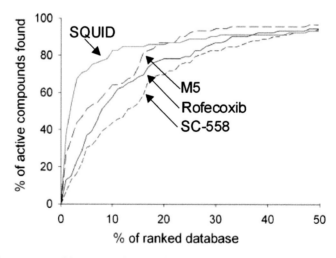

Figure 7. Enrichment curves of the retrospective screening with the SQUID model in comparison to the CATS3D descriptors of the COX-2 inhibitors used for the model calculation.

In Fig. 7 enrichment curves are shown from a comparison of SQUID fuzzy pharmacophores and CATS3D similarity searching for COX-2 ligands in the COBRA dataset of bioactive reference molecules [28]. Similarity searching was performed with the three COX-2 inhibitors from the alignment.

In this example SQUID fuzzy pharmacophore models outperformed similarity searching. While SQUID retrieved 75 % of the active molecules in the first 6 % of the ranked result-database, the best similarity search with rofecoxib resulted in 75 % of the actives ranked into the top 16 % of the database. This result demonstrates that it can be beneficial to integrate information from multiple molecules with the desired activity into a pharmacophore query rather than perform multiple individual database searches.

The combination of fuzziness and conservation of important features among the molecules provides a promising means for hopping from one "activity island" to another in chemical space.

ACKNOWLEDGEMENTS

The authors are most grateful to Dr Petra Schneider and Andreas Schüller for helpful discussions and their research contributions. Dr Lutz Weber is warmly thanked for leaving us the Ugi-data set to work with. This research was supported by the Beilstein-Institut zur Förderung der Chemischen Wissenschaften.

REFERENCES

[1]   Franz, M.O., Mallot, H.A. (2000) Biomimetic robot navigation. *Robot. Autonom. Syst.* **30**:133-153.

[2]   Todeschini, R., Consonni, V. (2000) *Handbook of Molecular Descriptors.* Wiley-VCH, Weinheim.

[3]   Byvatov, E., Fechner, U., Sadowski, J., Schneider, G. (2003) Comparison of support vector machines and artificial neural networks for drug-likeness prediction. *J. Chem. Inf. Comput. Sci.* **43**:1882-1889.

[4]   Willett, P. (2000) Chemoinformatics - similarity and diversity in chemical libraries. *Curr. Opin. Biotechnol.* **11**:85-88.

[5]   Dean, P.M., Lewis; R.A. Eds. (1999) *Molecular Diversity in Drug Design.* Kluwer Academic, Dordrecht.

[6]   Rarey, M., Stahl, M. (2001) Similarity searching in large combinatorial chemistry spaces. *J. Comput. Aided Mol. Des.* **15**:497-520.

[7]   Raymond, J.W., Willett, P. (2002) Effectiveness of graph-based and fingerprint-based similarity measures for virtual screening of 2D chemical structure databases. *J. Comput. Aided Mol. Des.* **16**:59-71.

[8]   Xue, L., Godden, J.W., Stahura, F.L., Bajorath, J. (2003) Design and evaluation of a molecular fingerprint involving the transformation of property descriptor values into a binary classification scheme. *J. Chem. Inf. Comput. Sci.* **43**:1151-1157.

[9]   Xue, L., Godden, J.W., Bajorath, J. (1999) Database searching for compounds with similar biological activity using short binary bit string representations of molecules. *J. Chem. Inf. Comput. Sci.* **39**:881-886.

[10]   Mason, J.S., Cheney, D.L. (1999) Ligand-receptor 3-D similarity studies using multiple 4-point pharmacophores. *Pac. Symp. Biocomput.* pp. 456-467.

[11]   Livingstone, D.J. (2000) The characterization of chemical structures using molecular properties. A survey. *J. Chem. Inf. Comput. Sci.* **40**:195-209.

[12]   Ghose, A.K., Viswanadhan, V.N., Wendoloski, J.J. (2001) Fundamentals of pharmacophore modeling in combinatorial chemistry. In: *Combinatorial Library Design and Evaluation.* (Ghose, A.K., Viswanadhan, V.N., Eds), pp. 51-71.Marcel Dekker, New York.

[13]   Schnitker, J., Gopalaswamy, R., Crippen, G.M. (1997) Objective models for steroid binding sites of human globulins. *J. Comput. Aided Mol. Des.* **11**:93-110.

[14]   Cui, S., Wang, X., Liu, S., Wang, L. (2003) Predicting toxicity of benzene derivatives by molecular hologram derived quantitative structure-activity relationships (QSARS). *SAR QSAR Env. Res.* **14**:223-231.

[15]   Jewell, N.E., Turner, D.B., Willett, P., Sexton, G.J. (2001) Automatic generation of alignments for 3D QSAR analyses. *J. Mol. Graph. Model.* **20**:111-121.

[16]   Broto, P., Moreau, G., Vandyke, C. (1984) Molecular structures: Perception, autocorrelation descriptor and SAR studies. *Eur. J. Med. Chem.* **19**:66-70.

[17]   Bauknecht, H., Zell, A., Bayer, H., Levi, P., Wagener, M., Sadowski, J., Gasteiger, J. (1996) Locating biologically active compounds in medium-sized heterogeneous datasets by topological autocorrelation vectors: dopamine and benzodiazepine agonists. *J. Chem. Inf. Comput. Sci.* **36**:1205-1213.

[18]   Anzali, S., Barnickel, G., Krug, M., Sadowski, J., Wagener, M., Gasteiger, J., Polanski, J. (1996) The comparison of geometric and electronic properties of molecular surfaces by neural networks: application to the analysis of corticosteroid-binding globulin activity of steroids. *J. Comput. Aided Mol. Des.* **10**:521-534.

[19]   Zupan, J., Gasteiger, J. (1999) *Neural Networks in Chemistry and Drug Design.* Wiley-VCH, Weinheim.

[20]   Schneider, G., Chomienne-Clement, O., Hilfiger, L., Kirsch, S., Böhm, H.J., Schneider, P., Neidhart, W. (2000) Virtual screening for bioactive molecules by evolutionary *de novo* design. *Angew. Chemie Int. Ed.* **39**:4130-4133.

[21]   Barnard, J.M., Downs, G.M., Willett, P. Descriptor-based similarity measures for screening chemical databases. In: *Virtual Screening for Bioactive Molecules.* (Böhm, H.J., Schneider, G., Eds), pp.59-80. Wiley-VCH, Weinheim.

[22]   Schneider, G., Nettekoven, M. (2003) Ligand-based combinatorial design of selective purinergic receptor A(2A) antagonists using self-organizing maps. *J. Comb. Chem.* **5**:233-237.

[23]   Schuffenhauer, A., Floersheim, P., Acklin, P., Jacoby, E. (2003) Similarity metrics for ligands reflecting the similarity of the target proteins. *J. Chem. Inf. Comput. Sci.* **43**:391-405.

[24]   Stahl, M., Rarey, M., Klebe, G. (2001) Screening of drug databases. In: *Bioinformatics: From Genomes to Drugs*. (Lengauer, T., Ed.), Vol. 2, pp.137-170. Wiley-VCH, Weinheim.

[25]   Johnson, M., Maggiora, G.M. (1990) *Concepts and Applications of Molecular Similarity*. John Wiley & Sons, New York.

[26]   Ginn, C.M.R., Willett, P., Bradshaw, J. (2000) Combination of molecular similarity measures using data fusion. *Perspec. Drug Discov. Des.* **20**:1-16.

[27]   Fechner, U., Franke, L., Renner, S., Schneider, P., Schneider, G. (2003) Comparison of correlation vector methods for ligand-based similarity searching. *J. Comput. Aided Mol. Des.* **17**:687-698.

[28]   Schneider, P., Schneider, G. (2003) Collection of bioactive reference compounds for focused library design. *QSAR Comb. Sci.* **22**:713-718.

[29]   Schneider, G., Neidhart, W., Giller, T., Schmid, G. (1999) "Scaffold-Hopping" by topological pharmacophore search: a contribution to virtual screening. *Angew. Chem. Int. Ed.* **38**:2894-2896.

[30]   MOE, Molecular Operating Environment. Distributor: Chemical Computing Group, 1010 Sherbrooke St. West, #910, Montreal, Canada H3A, http://www.chemcomp.com.

[31]   Bush, B.L., Sheridan, R.P. (1993) PATTY: A programmable atom typer and language for automatic classification of atoms in molecular databases. *J. Chem. Inf. Comput. Sci.* **33**:756-762.

[32]   Gasteiger, J., Rudolph, C., Sadowski, J. (1990) Automatic generation of 3D-atomic coordinates for organic molecules. *Tetrahedron Comput. Methodol.* **3**:537-547.

[33]   Fechner, U., Schneider, G. (2004) Evaluation of distance metrics for ligand based similarity searching. *Chembiochem* **5**:538-540.

[34]   Rarey, M., Kramer, B., Lengauer, T., Klebe, G. (1996) A fast flexible docking method using an incremental construction algorithm. *J. Mol. Biol.* **261**:470-489.

[35]   Jones, G., Willett, P., Glen, R.C., Leach, A.R., Taylor, R. (1997) Development and validation of a genetic algorithm for flexible docking. *J. Mol. Biol.* **267**:727-748.

[36]   Guner, O. (Ed.) (2000) *Pharmacophore Perception, Development and Use in Drug Design*. International University Line, La Jolla.

[37]   Pickett, S. (2003) The biophore concept. In: *Protein-Ligand Interactions*. (Böhm, H.J., Schneider, G., Eds), pp.72-105.Wiley-VHC, Weinheim.

[38]   Renner, S., Schneider, G. (2004) Fuzzy pharmacophore models from molecular alignments for correlation-vector-based virtual screening. *J. Med. Chem.* **47**:4653-4664.

[39]   Palomer, A., Cabre, F., Pascual, J., Campos, J., Trujillo, M., Entrena, A., Gallo, M., Garcia, L., Mauleon, D., Espinosa, A. (2002) Identification of novel cyclooxygenase-2 selective inhibitors using pharmacophore models. *J. Med. Chem.* **45**:1402-1411.

EVOLUTIONARY PERSPECTIVES ON PROTEIN STABILITY

RICHARD A. GOLDSTEIN

Division of Mathematical Biology, National Institute for Medical Research, Mill Hill, London, NW7 1AA, U.K.

E-Mail: richard.goldstein@nimr.mrc.ac.uk

Received: 6th September 2004 / Published: 22nd July 2005

ABSTRACT

Biochemists, molecular biologists, and biophysicists, when confronted with the exquisite matching of natural organisms to their environment, have generally interpreted the properties of proteins as resulting from adaptation. Conversely, since 1962, evolutionary theory has been emphasizing the stochastic nature of neutral evolution and random drift. It is actually possible to explain many of the seemingly adaptive features of proteins as resulting through neutral evolution. In this paper, we use a simple computational model to demonstrate how marginal stability as well as the robustness of proteins to site mutations can be explained by neutral evolution of populations.

INTRODUCTION

If we were to analyse a computer looking only at its constituent parts, we might be able to characterize the physical properties of the case, the chemical properties of the circuit boards, and the ferromagnetic properties of the rotating disks. We would, however, be unable to understand why the components were the way they are, and thus would have trouble understanding how they worked together to give the collective behaviour observed with computers. This is because we had ignored the *functional* description, the discussion of the role and purpose for each of these components. This functional definition is appropriate because the computer is a result of a design process, where teams of engineers have created the computer as a tool for specific purposes, and the components reflect their use in these tasks. Conversely, it would be quite difficult to analyse the workings of a computer without any information about the physical, chemical, etc., properties of the components.

We could not understand how the various parts of the tasks were organized or implemented without knowing the opportunities, constraints, and limitations of the materials and processes available to the engineers. An understanding of a modern computer then requires both a description of the properties of the components in the context of the roles for which the computer was designed.

There has not been a corresponding 'design process' for living systems, no teams of engineers that have drawn schematics for how the cell should be constructed. Yet there is an overarching biological process that in some ways can resemble engineering design. This is, of course, the process of evolution, where living systems are evaluated based on their ability to survive and reproduce, with the 'winners' continuing into the next round. For this reason, we can ask 'why' questions: what is the 'reason' for the existence of an organ, tissue, cell, enzyme, etc. Understanding how these components have developed so as to contribute to survival and reproduction can give us many insights into their role and purpose, and how they work together to fulfil the required tasks. We cannot neglect, however, the opportunities, constraints, and limitations imposed on the process of evolution based on the physical and chemical properties of the components. By looking at biological systems in this evolutionary context, we can combine these two perspectives: understanding how evolution adapts the various components to the needs of survival and reproduction, while being constrained by the (constantly changing) properties of these components. This is in some ways the essence of the Chemical Theatre of Biological Systems: how the chemical properties of molecular systems determine the biological properties of living systems, while the biological process of evolution underlies the resulting chemical properties of the biological components.

It is important to keep in mind how evolution differs from engineering design. While randomness occurs in engineering work (a designer reads an inspiring article in a magazine found in the train), it is a central and indispensable aspect of evolution. Evolution can only work with the raw materials placed in its hands, the variations that occur at each generation, variations that result from the stochastic chemical reactions that produce site mutations and the random process of recombination, gene conversion, chromosome sorting, etc. All of this is constrained by the need for each intermediate individual to be capable of reproduction, for there to be an acceptable path from one genotype to another, viable at every point in between.

Selection, the mechanism, which governs the number of offspring an individual contributes to the next generation, is also extremely fickle. Favourable mutations can be eliminated, deleterious mutations can become fixed.

It is impossible to look at living systems without being amazed at the efficiency and power of evolution, how exquisitely well adapted such systems are. As a result, evolutionary thinking has often concentrated on the role of 'positive adaptation', the process by which increasingly fit members of the population are generated and selected embodied by the phrase, 'survival of the fittest'. There are dangers to neglecting evolutions' stochastic nature, that we can fall into what Gould calls the 'Panglossian paradigm', analysing how all is for the best in this 'best of all possible worlds' [1]. In contrast to the models that emphasize the role of adaptation, the 'neutral theory' developed in the 1960s [2-4] focuses on the consequences of the random processes of variation and drift. According to this theory, while an adaptive mutation is more likely to be accepted in a population, all but a small fraction of mutations are deleterious or neutral, all with corresponding probabilities of acceptance. As a result, most accepted changes are actually neutral or slightly deleterious. The stochastic nature of these changes does not preclude the derivation of important principles and concepts, any more than the stochastic nature of the motion of gas particles invalidates the ideal gas law. We need to use a different type of thinking, of looking at the evolutionary equivalents for concepts such as entropy. While evolutionary biology has been debating the importance and consequences of adaptation and neutral drift, biochemistry and molecular biology has generally analysed biological systems in terms of a purely adaptionist model. Little has been done looking at the role of neutral drift in determining the nature of biological systems at a molecular level.

In this paper I look at how proteins can be considered in an evolutionary context, in particular, in how much can be explained by the process of neutral evolution. While obviously adaptation occurs, such a perspective offers a number of advantages. Firstly, because neutral drift is always taking place, a demonstration that a characteristic of proteins can result from neutral drift provides evidence that neutral evolution is a *sufficient* explanation, and is in fact a more *parsimonious* explanation, that should only be rejected when it is shown to be clearly inadequate. Secondly, considering neutral drift protects us against the 'Panglossian paradigm'. Thirdly, failure to be able to explain a phenomenon based on neutral evolution highlights those aspects that may require an adaptionist explanation. And lastly, such a perspective helps to remind us that evolution is *not* an optimization procedure.

Populations change and vary according to stochastic, discrete differential equations oblivious to any terms such as 'fitness'. There is no teleological underpinning to biological evolution except that imposed by the observer for their own intellectual convenience, or our psychological need to imagine ourselves at the peak of perfection in *some* space.

Proteins are a convenient choice for such investigations, in that they span a simplified form of the genotype/phenotype divide. Evolution involves the changing of the sequential information contained in the genotype as represented by the composition of biomolecules, generally DNA. Selection, however, acts on the resulting phenotypic traits representing the *interpretation* of the genotype through the process of development. Proteins can be characterized by their amino acid sequence (a close reflection of the genotype), as well as by their physical and chemical properties, a simplified form of phenotype. In this way they can provide insight into the evolution of higher organizational forms. In addition, proteins are important and interesting in their own right. They are involved in essentially every biological process, are basic to understanding all of these processes at a molecular level, and form the targets of most pharmaceutical products. Altered, 'engineered' proteins have the potential to fulfil many different tasks for us. Various aspects of proteins form the basis for many pathological conditions. Many processes involving proteins (such as protein folding) are still poorly understood.

Proteins are still extremely complicated. They are large molecules which involve a number of different types of atoms and functional groups, interacting with solvent, membranes, ions, etc. They have many characteristics, including structural, functional, kinetic and thermodynamic properties. They exist in a complicated biological context. To analyse the general principles of a realistic model of protein evolution is beyond our capabilities. One approach is to use realistic models but to focus our attention to more manageable properties of specific proteins in specific situations. Alternatively, we can use highly simplified models, restricting our inquiries to questions for which these models are appropriate, making all possible connections with experimental observations. We choose the latter approach here. In particular, we represent proteins as two-dimensional lattice models. We focus our attention on the stability of proteins, both to thermal fluctuations and mutations.

MODELS

The model used for these calculations involves a) the model of the evolving proteins, and b) the model used for the evolutionary behaviour, explicitly including population effects.

Models of the Proteins

Figure 1 shows the model of the protein used. Proteins are represented as 25 residue polypeptides confined to a 5 × 5 square lattice, where each residue occupies one lattice point.

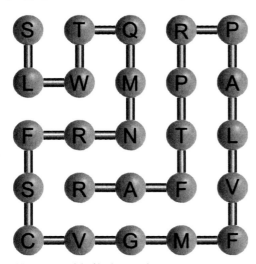

The 1081 possible conformations for the protein represent all self-avoiding walks through the lattice, not counting rotations, reflections and inversions. The two-dimensional lattice is obviously a highly simplified representation, and would be inappropriate for dynamics studies (where the set of conformations are generally non-ergodic, depending upon the move-set [5]) but this model does allow us to have a reasonable surface-to-volume ratio with a moderately-sized protein.

Figure 1. Model of lattice protein.

The energy $E(S,C)$ of a given sequence S in conformation C is a pairwise contact energy, given by

$$E(S,C) = \sum_{<i,j>} \gamma(A_i, A_j)\, u_{i,j} \tag{1}$$

where $\gamma(A_i, A_j)$ is the interaction energy between the amino acid types located at locations i and j in the sequence (such as between the serine and threonine in the upper-left corner of the protein shown in Fig. 1) and $u_{i,j}$ is 1 if residues i and j are in contact (that is, are not covalently connected but are on nearest neighbouring lattice points), and zero otherwise. We used the values of $\gamma(A_i, A_j)$ calculated by Miyazawa and Jernigan based on their statistical analysis of a database of known protein structures [6]. These interaction terms represent 'potentials of mean force' that implicitly include the interaction of the residues with the solvent.

We assume, based on earlier work, that the conformation of lowest energy represents the native state C_{ns} [7]. The probability P_{ns} that a protein at equilibrium at temperature T would be in this state is

$$P_{ns} = \frac{\exp(E(S,C_{ns})/kT)}{\sum_C \exp(E(S,C)/kT)} \tag{2}$$

where the lower sum is over all conformations. This allows us to compute $\Delta G_{folding}$

$$\Delta G_{folding} = -kT \log\left(\frac{P_{ns}}{1-P_{ns}}\right) \tag{3}$$

$$= -kT \log\left(\frac{\exp(E(S,C_{ns})/kT)}{\sum_{C \neq C_{ns}} \exp(E(S,C_{ns})/kT)}\right) \tag{4}$$

The interaction parameters are in kcal/mol, so all free energies are in these units.

Modelling Population Evolution

Population dynamics form a central element in the evolutionary process. In particular, it is the fact that populations are finite (and much smaller than, for instance, Avogadro's number) that introduces the large stochastic element. For these reasons, we include such dynamics in our simulations. In general, we start with a population of identical protein sequences. Given a specified average mutation rate, we select the number of mutations in each particular generation with the approximate Poisson distribution. These residues as well as the proteins in which they belong are selected at random, and each amino acid is changed at random to one of the other 19. The various properties of the resulting proteins are computed, including the ground state and the thermal stability. We then eliminate 'non-viable' sequences, the sequences that fail to meet some appropriately chosen selection criterion, such as minimum acceptable thermostability. In order to represent the random process or reproduction, we choose new sequences at random from the set of surviving sequences from the previous generation, with replacement, until we have the same number of sequences as we had initially. This represents the population at the next generation. The population size remains fixed during the simulation.

RESULTS

Protein Thermostability

It has been long noted that proteins are marginally stable, with stabilities ($-\Delta G_{folding}$) typically observed of around 10 kcal/mol. The most general explanation for this effect is that proteins have evolved to be marginally stable, that marginally stable proteins are more 'fit', that marginal stability represents an adaptation. A number of different reasons have been given why marginal stability would be advantageous. For instance, protein functionality might require a degree of flexibility incompatible with high stability [8,9] Alternatively, protein flexibility would destabilize ligand binding, allowing greater control and better abilities to modulate the affinity through mutation, post-translational modification, or other slight changes in chemical properties [10-12]. Finally, marginal stability might increase the ability to degrade proteins, an important aspect of cellular control. An alternative explanation for the observed marginal stability is that this involves 'optimization given constraints', that proteins have to fulfil so many other selective criteria involving functionality, rigidity, solubility, etc., that proteins can only increase their stability to a certain point without compromising these other factors. This second explanation is also adaptionist in nature, but emphasizes the more complicated nature of fitness, and that all components to the fitness cannot be independently optimized.

Both of these explanations - that proteins that are marginally stable are more fit, or that stability is optimised given constraints - are adaptionist in nature. What would occur during neutral evolution, where neither of these mechanisms is active? To answer this question, we modelled a population of 3000 sequences, collecting data for 30,000 generations following an initial 30,000 generations for equilibration ('burn-in') [13]. The mutation rate was maintained at a rate of 0.2% mutations per protein per generation. Proteins were considered viable if they were 'adequately' stable, that is, with a ΔG of folding $\Delta G_{folding}$ less than some 'critical' ΔG_{crit}.

The results for $\Delta G_{crit} = 0$ is shown in Fig. 2A. There are many extremely stable sequences that can be obtained through hill-climbing methods, but the evolutionary runs generally result in marginally-stable proteins. In these simulations, there were no constraints placed on the optimization, nor were there any advantages to marginal stability. Proteins are naturally marginally stable because the vast number of viable sequences ($\Delta G_{folding} < \Delta G_{crit}$) is marginally stable ($\Delta G_{folding} \approx \Delta G_{crit}$).

This can be seen by comparison with the distribution observed for random sequences with $\Delta G_{folding} < \Delta G_{crit}$, also shown in Fig. 2A. Again, the vast proportion of such sequences has $\Delta G_{folding} \approx \Delta G_{crit}$. This is because they represent the extreme tail of a broader distribution of random stabilities, as shown in Fig. 2B.

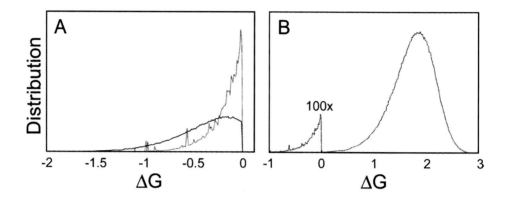

Figure 2. A) Distribution of stabilities observed for the proteins resulting from population dynamics with $\Delta G_{folding} < \Delta G_{crit} = 0$ (solid line), compared with the distribution for randomly selected proteins, also with $\Delta G_{folding} < 0$ (dashed line). **B)** The overall distribution of stabilities of random proteins.

One way to make intuitive sense of these results is to consider the abstract space of all possible sequences, represented in Fig. 3. This space is extremely high-dimensional (as many dimensions as the length of the sequences), but also extremely sparse (only 20 points along each dimension). The vast majority of this space consists of non-viable sequences, sequences that would not fold or would not be stable in any given fold. There are sequences in various regions of the space, shown in grey, which would be able to fold into one of a number of possible structures. One aspect of high dimensional spaces is the surface to volume ratio: the fraction of the volume of a hypersphere of dimension n and radius r occupied by a thin shell of radius δr goes as $\dfrac{n\,\delta r}{r}$, meaning that for a 100-dimensional space (corresponding to a 100-residue protein), 99% of the sequence volume would be in the outer 1% of the hypersphere. If the region inside the hypersphere represents viable proteins - foldable and stable - and the region outside the hypersphere represents non-viable proteins - unfoldable and unstable - this suggests that the vast majority of viable proteins are barely viable.

This is purely the result of the high dimensionality of the sequence space, assuming that the viability is slowly varying (so regions on the edge of the viable region are barely viable), and that the regions in sequence space corresponding to viable proteins form relatively compact objects.

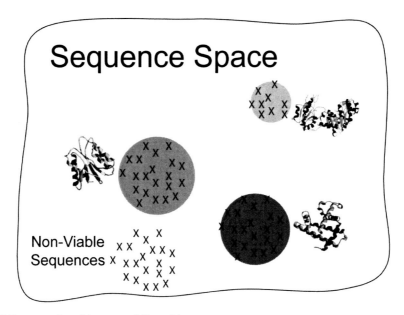

Figure 3. Representation of the space of all possible sequences.

In situations such as these, if the vast majority of viable protein sequences are marginally stable, there is no need to propose other evolutionary mechanisms (such as adaptation) to account for this observation in biological proteins. Neutral evolution is a sufficient, and therefore more parsimonious, explanation. In this case, the tendency for proteins to be marginally stable represents 'sequence entropy': in a random walk, the majority of sequences visited will mirror the location of the majority of random viable sequences. Sequence entropy can drive proteins towards marginal stability.

It has been observed that proteins that have modified in order to increase their thermostability sometimes display lower than native function [14-17]. Does this indicate that proteins have evolved for marginal stability in order to increase function? Firstly, the result is controversial. But even if the observation were correct and general, we can still analyse it through the Panglossian paradigm. According to Professor Pangloss, a character in Voltaire's Candide,

"…our nose was made to carry spectacles, so we have spectacles.

Legs were clearly intended for breeches, and we wear them." [18].

Removing our noses would greatly affect our ability to wear spectacles. Yet it still does not provide any evidence that this is actually what our noses are 'for'. Or to use another analogy, globular proteins have evolved to function in the aqueous environment of the cytosol. One with less understanding of evolutionary biology might incorrectly conclude that our cytosol is aqueous in order to provide the best environment for globular proteins. And to provide more evidence for this claim, they might note that these proteins do not function well in non-aqueous environments. But of course proteins have evolved in a certain context, and changing this context can change the ability of the proteins to function, regardless of whether the context preceded the protein evolution. Proteins evolved in the context of a highly-dimensional sequence space where marginal stability is most likely, and the various properties that emerged during evolution were compatible with this context.

We can see this more directly in an evolutionary competition between different sets of proteins [13]. Consider three different types of proteins, which fulfil the same function through three different mechanisms. One type of protein can function if it has high stability ($\Delta G_{folding} < -2$), one type of protein requires moderate stability to function ($-2 < \Delta G_{folding} < -1$), and the third type requires marginal stability ($-1 < \Delta G_{folding} < 0$). A highly stable protein of the first type has identical fitness to a moderately stable protein of the second type, which has the identical fitness as a marginally stable protein of the third type. Any protein that mutated into a sequence with a value of $\Delta G_{folding}$ not within its functional range would be considered non-viable and eliminated from the simulation, as described above. Three different populations, one of each type, were allowed to equilibrate separately, and then mixed together in a single evolutionary run. In 24 out of 25 runs, the proteins that required marginal stability to function dominated the population by the end. (One time the protein type that required moderate stability became dominant, emphasizing the stochastic nature of evolution.) This was not because there was any positive selection, or any selective advantage at all for the proteins that required marginal stability. It was just that the functionality arose that was most consistent with the type of proteins that are formed naturally during the runs, those that were marginally stable. In the non-intuitive world of evolution, proteins evolved to require marginal stability because they were naturally marginally stable!

Again, it is important to emphasize that these simulations do not demonstrate that marginal stability results from neutral evolution. Only that it would result from neutral evolution, this is the most parsimonious explanation, and there is no need to look for an alternative explanation. The observation of marginal stability does not provide any evidence for any selective pressure for marginal stability.

Evolutionary Robustness

An alternative method to analyse the tendency for our evolutionary model to favour functional mechanisms consistent with marginal stability is to consider the role of mutational robustness. Fitness is related to the number of viable offspring likely in the next generation. A fully viable protein that requires high stability is quite likely to have mutated offspring that do not have the requisite stability. Proteins with marginal stability are more likely to have their mutated offspring having the marginal stability that they need for function. Proteins with functions that require high stability wander in a much more treacherous fitness landscape.

This can be seen directly in what has been described as 'the survival of the flattest' [19]. Consider a set of random protein sequences. For each sequence, we can calculate its initial stability, which we will refer to as ΔG_{wt}. We can also look at random mutations of this 'wild type' sequence. In general mutations will change the stability. Figure 4 shows the probability that a mutation is destabilizing, that is, $\Delta\Delta G_{folding} > 0$ [20].

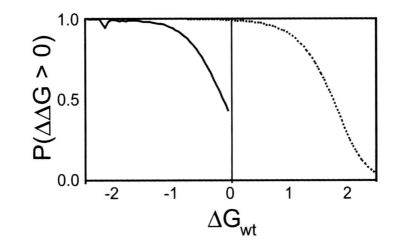

Figure 4. Probability a site mutation is destabilising as a function of the initial stability, for evolved proteins ($\Delta G_{crit} = 0$, solid line) compared with random proteins (dashed line).

We now look at a similar quantity for the proteins that result from population dynamics, where the proteins have evolved so that they survive only if $\Delta G_{folding} < 0$. If we take these sequences and make new, random mutations, we find that the probability of a destabilizing mutation, for a given value of ΔG_{wt}, is much lower. This means that two sequences with identical values of $\Delta G_{folding}$, possibly the same structure, one derived from a population simulation, the other chosen at random, can have vastly different responses to site mutations.

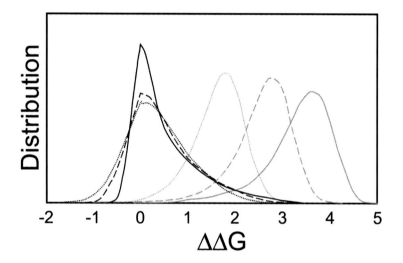

Figure 5. Distribution of changes in stability ($\Delta\Delta G$) for evolved proteins (black lines; $\Delta G_{crit} = 0$ (short dashes), -1 (long dashes), and -2 (solid)), compared with the distribution for random proteins (grey lines; $\Delta G_{folding} < 0$ (short dashes), -1 (long dashes), and -2 (solid)).

Figure 5 demonstrates this robustness in a different manner. In this figure, we show the distribution of $\Delta\Delta G$ values for random sequences (chosen with $\Delta G_{folding} < 0$, -1, and -2), compared with the corresponding distribution for evolved sequences (evolved with $\Delta G_{crit} = 0$, -1, and -2). For the random sequences, the more stable they are, the more destabilizing the average mutation. In fact, for every unit of increased initial stability, there is about an extra unit decrease in stability resulting from a random mutation. Conversely, for evolved sequences, the more stable the proteins, the more likely it is that the mutation will have little or no effect on the stability! Again, population dynamics work with evolution in order to select sequences that have fit offspring, that is, have robustness to mutations.

An alternative perspective can be obtained by considering the work of Eigen on biomolecular evolution [21]. Individuals do not evolve; populations do, representing a distribution of genotypes and their resultant phenotypes, what Eigen called a 'pseudospecies'. This distribution is maintained by the centrifugal nature of mutations. Rather than considering the fitness of a single wild-type sequence, we need to consider the averaged fitness over the entire distribution. A lower, flatter region of the fitness landscape might actually have a higher average fitness than a higher, more sharply peaked region. The population dynamics result in the selection for a network of related sequences that share a general level of fitness, and thus are robust to mutations through this network.

This perceived robustness of proteins to substitutions has been observed experimentally. For instance, Reddy *et al.*, catalogued a wide range of mutations, observing that approximately 25% actually increased thermal stability [22].

CONCLUSION

In contrast to the general tendency for biochemists and molecular biologists to imagine evolution as a march up the fitness landscape to higher and higher levels, modern evolutionary biology has emphasized the importance of evolution's stochastic nature, and the consequences of neutral evolution. While biochemists and molecular biologists are likely to try to explain an observed characteristic of living systems by asking how it serves to increase the fitness of those that possess it, many of these properties can be explained by neutral drift interacting with population dynamics. In this manner, sequence entropy - the number of sequences with a given characteristic - becomes important. Sequence entropy has been used to explain why some substructures and structures are over-represented in biological proteins [23-27] why proteins might fold to the structure of lowest free energy [7] and in work summarized here, why proteins are naturally marginally stable [13,28] and robust to site mutations [20].

One consequence of this perspective relates to the opportunities available in protein engineering. The work described above makes the prediction that biologically-derived proteins would be distinct from random sequences with similar properties by being especially robust to mutations. This suggests that it would be possible to modify naturally-occurring proteins to develop novel properties, and that other important qualities - structural rigidity, thermostability, etc. - would be maintained.

Conversely, the tendency of evolution to select flatter regions of the fitness landscape, ignoring potentially higher, narrower, peaks, means that there may be higher fitness peaks available to those who wish to find them through non-evolutionary means.

ACKNOWLEDGEMENTS

The work described here was largely performed by Sridhar Govindarajan, Darin Taverna, and Paul Williams.

REFERENCES

[1]   Gould, S.J., Lewontin, R.C. (1979) The spandrels of San Marco and the Panglossian paradigm: A critique of the adaptationist programme. *Proc. R. Soc. Lond. B* **205**:581-598.

[2]   Sueoka, N. (1962) On the genetic basis of variation and heterogeneity of DNA base composition. *Proc. Natl. Acad. Sci. USA* **48**:582-592.

[3]   Kimura, M. (1968) Evolutionary rate at the molecular level. *Nature* (Lond.) **217**:624-626.

[4]   King, J.L., Jukes, T.H. (1969) Non-Darwinian evolution. *Science* **164**:788-798.

[5]   Abkevich, V.I., Gutin, A.M., Shakhnovich, E.I. (1995) Impact of local and non-local interactions on thermodynamics and kinetics of protein folding. *J. Mol. Biol.* **252**:460-471.

[6]   Miyazawa, S., Jernigan, R.L. (1985) Estimation of effective interresidue contact energies from protein crystal structures: Quasi-chemical approximation. *Macromolecules* **18**:534-552.

[7]   Govindarajan, S., Goldstein, R.A. (1998) On the thermodynamic hypothesis of protein folding. *Proc. Natl. Acad. Sci. USA* **95**:5545-5549.

[8]   Wagner, G., Wuthrich, K. (1979) Correlation between the amide proton exchange rates and the denaturation temperatures in globular proteins related to the basic pancreatic trypsin inhibitor. *J. Mol. Biol.* **130**:31-37.

[9]   Tang, K.E.S., Dill, K.A. (1998) Native protein fluctuations: The conformational-motion temperature and the inverse correlation of protein flexibility with protein stability. *J. Biomol. Struct. Dyn.* **16**:397-411.

[10]   Dunker, A.K., Garner, E., Guilliot, S., Romero, P., Albrecht, K., Hart, J., Obradovic, Z., Kissinger, C., Villafranca, J.E. (1998) Protein disorder and the evolution of molecular recognition: Theory, predictions and observations. *Pacific Symp. Biocomputing* **3**:473-484.

[11] Wright, P.E., Dyson, H.J. (1999) Intrinsically unstructured proteins: Re-assessing the protein structure-function paradigm. *J. Mol. Biol.* **293**:321-331.

[12] Dunker, A.K., Obradovic, Z. (2001) The protein trinity-linking function and disorder. *Nat. Biotechnol.* **19**:805-806.

[13] Taverna, D.M., Goldstein, R.A. (2002) Why are proteins marginally stable? *Proteins: Struct. Funct. Genet.* **46**:105-109.

[14] Alber, T., Wozniak, J.A. (1985) A genetic screen for mutations that increase the thermal-stability of phage-T4 lysozyme. *Proc. Natl. Acad. Sci. USA* **82**:747-750.

[15] Bryan, P.N., Rollence, M.L., Pantoliano, M.W., Wood, J., Finzel, B.C., Gilliland, G.L., Howard, A.J., Poulos, T.L. (1986) Proteases of enhanced stability: Characterization of a thermostable variant of subtilisin. *Proteins: Struct. Funct. Genet.* **1**:326-334.

[16] Liao, H., McKenzie, T., Hageman, R. (1986) Isolation of a thermostable enzyme variant by cloning and selection in a thermophile. *Proc. Natl. Acad. Sci. USA* **83**:576-580.

[17] Shoichet, B. K., Baase, W. A., Kuroki, R., Matthews, B. W. (1995) A relationship between protein stability and protein function. *Proc. Natl. Acad. Sci. USA* **92**:452-456.

[18] Voltaire, F.M. (1759) (real name Arouet, F. M.), *Candide*, (John Butt, trans.) Penguin, London.

[19] Nimwegen, E. v., Crutchfield, J.P., Huynes, M. (1999) Neutral evolution of mutational robustness. *Proc. Natl. Acad. Sci. USA* **96**:9716-9720.

[20] Taverna, D.M., Goldstein, R.A. (2002b) Why are proteins so robust to site mutations? *J. Mol. Biol.* **315**:479-484.

[21] Eigen, M. (1971). Self organization of matter and the evolution of biological macromolecules. *Naturwissenschaften* **10**:465-523.

[22] Reddy, B.V.B., Datta, S., Tiwari, S. (1998) Use of propensities of amino acids to the local structure environment to understand effect of substitution mutations on protein stability. *Protein Engng* **11**:1137-1145

[23] Finkelstein, A.V., Ptitsyn, O.B. (1987) Why do globular proteins fit the limited set of folding patterns. *Prog. Biophys. Mol. Biol.* **50**:171-190.

[24] Govindarajan, S., Goldstein, R.A. (1995) Searching for foldable protein structures using optimized energy functions. *Biopolymers* **36**:43-51.

[25] Govindarajan, S., Goldstein, R.A. (1996) Why are some protein structures so common? *Proc. Natl. Acad. Sci. USA* **93**: 3341-3345.

[26] Li, H., Helling, R., Tang, C., Wingreen, N. (1996) Emergence of preferred structures in a simple model of protein folding. *Science* **273**:666-669.

[27] Shakhnovich, E.I. (1998) Protein design: A perspective from simple tractable models. *Folding & Design* **3**:R45-R58.

[28] Williams, P.D., Pollock, D.D., Goldstein, R.A. (2001) Evolution of functionality in lattice proteins. *J. Mol. Graphics Modell.* **19**:150-156.

235

MODELLING AND SIMULATION OF PHARMACOKINETIC AND PHARMACODYNAMIC SYSTEMS - APPROACHES IN DRUG DISCOVERY

ALEX J. MacDONALD*, NEIL PARROTT, HANNAH JONES AND THIERRY LAVÉ

Modelling and Simulation, Pharma Research, F. Hoffmann-La Roche Ltd, Grenzacherstrasse, CH-4070 Basel, Switzerland

E-Mail: *alexander.macdonald@roche.com

Received: 30th June 2004 / Published: 22nd July 2005

ABSTRACT

Pharmacokinetics and pharmacodynamics in drug discovery are often viewed as simple data generation processes. Candidate compounds are screened for various ADME and physico-chemical properties together with their potency *in vitro* and their effectiveness in *in vivo* models. In many cases simple summary parameters are used for comparison between drug candidates and for project decision-making. However, weighing the relevance and importance of such data in isolation or in a qualitative manner is not a simple task. Modelling and simulation provides a framework for integrating these data, providing outputs that contain more information than can be elucidated from the data in isolation. The use of biologically realistic models allows for the separation of the biological and compound-specific components of the pharmacokinetic and pharmacodynamic systems. One can then begin to develop a generic approach that is applicable to the drug discovery process. Physiologically based pharmacokinetic (PBPK) modelling is integral to Roche's approach. PBPK models, by design, are capable of integrating information about various pharmacokinetic processes, including absorption, metabolism and distribution. They can be used not only to estimate summary *in vivo* pharmacokinetic parameters, but also to predict the complete drug concentration time-course in both plasma and tissues. However, a commonly held view is that PBPK models are complex and data-intensive, and therefore, not applicable to the early phases of drug development. Many of the biochemical and physico-chemical parameters are generated routinely *in vitro* in the lead generation and optimization phases.

Quantitative structure-activity relationships and mechanism-based *in silico* models exist for estimating other drug properties, most critically, the partitioning of drugs into different body tissues. *In vitro - in vivo* scaling methods are becoming routine for the estimation of hepatic metabolism. The necessary physiological and anatomical data (e.g. tissue volumes and blood flows) are available in the scientific literature for many commonly used laboratory animals and humans. In short, most of required data are already available.

By combining PBPK models with simple pharmacodynamic models, for example based on *in vitro* or *in vivo* efficacy data, the link between basic compound properties and effect *in vivo* is made. This allows the project teams to compare compounds with the target profile and with each other over a range of simulated doses and regimens. Crucially, an attempt at predicting drug effect *in vivo* in the target species, human, can be made long before the drug reaches the clinic.

INTRODUCTION

Advances in technology in drug discovery have improved the process of identifying biologically active compounds. However, choosing potential medicines from these compounds is difficult, and the pharmaceutical industry is still plagued with high failure rates at all stages of the drug development process [1]. Historically a high proportion of these failures have been due to poor pharmacokinetic properties. Candidate compounds are now routinely screened for their ADME (absorption, distribution, metabolism and excretion) and physico-chemical properties, in the attempt to discriminate between medicine-like and non-medicine like compounds. These data, together with *in vitro* and animal pharmacology data, are used to make both comparative and absolute assessments about whether compounds are likely to be effective and safe in the target population. Such assessments are essential from both an ethical and business perspective, allowing resources to be allocated to drug candidates which are likely to be both effective and safe. However, the ultimate integration and use of these data is a complex and uncertain task. Fortunately, modelling and simulation, utilizing the principles of pharmacokinetics and pharmacodynamics, provides a natural and rational framework for integrating these data. Such tools provide outputs that contain more information than can be elucidated from the data in isolation and allow for clear and consistent management of unknowns and uncertainties. By using biologically realistic models and methodologies, one can separate the pharmacokinetic and pharmacodynamic characteristics of a drug candidate into compound and biology-specific components.

The main advantage of such an approach is that one makes use of prior knowledge, for example animal and human physiological and anatomical data, to reduce the unknowns in an effort to minimize uncertainty and increase predictivity. In the following sections pharmacokinetic and pharmacodynamic systems will be discussed in terms of biologically-based mathematical modelling and their application in drug discovery.

PHARMACOKINETICS AND PHARMACODYNAMIC SYSTEMS

Pharmacokinetics (PK) is the study of the time-course of a drug in the body. Correspondingly, pharmacodynamics (PD) is the study of the time-course of drug action and is intrinsically linked to pharmacokinetics [2]. Figure 1 illustrates the concepts and the interdependence.

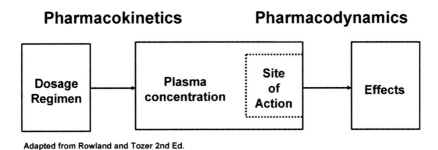

Adapted from Rowland and Tozer 2nd Ed.

Figure 1. Illustration of the dose-response relationship of a drug in terms of pharmacokinetics and pharmacodynamics.

As can be seen from Fig. 1, by applying the principles of pharmacokinetics and pharmacodynamics the dose-effect relationship for a drug is defined. Implicit within this approach is the use of mathematical models to approximate the pharmacokinetic and pharmacodynamic systems for a drug or drug-candidate. For the drug discovery scientist the interest in PK/PD modelling and simulation lies in its ability to estimate the likely dose-response of drug-candidates in human patients from limited animal and *in vitro* data. If these methods can be shown to be consistently predictive the impact on the drug development process in terms of both attrition rate and resources will be high. Approaches which are being developed and applied by Roche, in the attempt to predict human dose-response in the discovery stage of drug development, will be discussed in the following sections.

Pharmacokinetic Modelling

In order to estimate the dose-response relationship in the target population one first requires an estimate of the pharmacokinetics in the target population. The approaches used to predict human pharmacokinetics tend to fall into two categories: empirical interspecies scaling, based primarily on animal pharmacokinetic data and physiologically-based pharmacokinetic (PBPK) modelling [3-7]. With the recent developments of in silico and in vitro tools together with a marked increase in computing power, PBPK modelling is rapidly becoming a powerful tool for predicting human pharmacokinetics. Therefore, the focus of this paper is on physiological models and the reader is referred to the literature for further discussion on animal scaling methods [8-11].

Although as the title of this paper implies, pharmacokinetics should be viewed as a system of interlinked processes, it is often convenient to sub-divide the system into simpler components for the purposes of discussion. Therefore, PBPK modelling will be discussed initially in the context of absorption, distribution, metabolism and excretion processes.

Absorption

Oral absorption is determined by complex mechanisms which are governed by physiological and biochemical processes (e.g. pH in the various sections of the gut, gastric emptying, intestinal transit, active transport and intestinal metabolism), drug-specific properties (e.g. lipophilicity, pK_a, solubility, particle size, permeability, and metabolism) and formulation factors (e.g. release kinetics, dissolution kinetics). These are some of the main determinants which should be an integral part of a biologically-based absorption model. The interplay of parameters describing these processes determines the rate and extent of absorption. The available simulation tools to predict oral absorption in animals and humans have been reviewed recently [12-14]. Different absorption models have been developed and in part described in the literature. These physiological gut absorption models are developed to a degree that they are commercially available as software tools e.g. GASTROPLUS® from Simulations Plus Inc. (see Fig. 2). In brief, these models are physiologically based transit models segmenting the gut into different compartments, where the kinetics of transit, dissolution and uptake are described by sets of differential equations. The simulation models for oral absorption use a variety of measured or calculated *in vitro* input data such as permeability, solubility, pK_a and dose.

With the provided *in vitro* data, the rate of absorption into the portal vein is often only estimated as an intermediate step, from which the cumulative extent of absorption is derived as a point estimated for a given drug candidate and/or formulation.

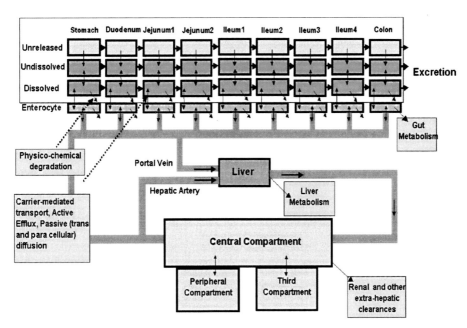

Figure 2. Physiological Gut Absorption Model GASTROPLUSTM (with permission Simulations Plus Inc.).

Distribution

The distribution and disposition of a drug can be determined by using the principles of mass conservation [5]. If an organ or tissue is considered as a set of compartments representing the volumes of the different fluid spaces within the tissue, the rate of change of drug mass in the tissue can be described by mass-balance ordinary differential equations. A general kinetic model of a tissue or organ is shown in Fig. 3A. The tissue is split into three compartments representing blood in the tissue, the interstitial space (IS) and intracellular space (IC), respectively. The interface between the blood and interstitial space compartments represents the blood vessel walls and the interface between interstitial and intracellular spaces represents cell membranes. The rate of change of amount of drug in tissue blood is:

$$\frac{dA_{VT}}{dt} = Q_T.C_A - Q_T.C_{VT} - FLUX_{P,IS}$$

where A_{VT} is the mass of drug in plasma, Q_T is the tissue blood flow, C_A is the concentration of drug entering the tissue in arterial blood, C_{VT} is the concentration of drug in effluent venous blood and $FLUX_{P,IS}$ is the net rate of movement of drug across the capillary walls. The rate of change of amount of drug in the interstitial space is:

$$\frac{dA_{IS}}{dt} = FLUX_{P,IS} - FLUX_{IS,IC}$$

where A_{IS} is the mass of drug in the interstitial space and $FLUX_{IS,IC}$ is the net rate of movement of drug across the tissue cell membrane. Correspondingly, the rate of change of amount of drug in the intracellular space is:

$$\frac{dA_{IC}}{dt} = FLUX_{IS,IC}$$

where A_{IC} is the mass of drug in the intracellular space. These models assume that there are no metabolism or excretion processes in any compartments. The flux or rate of transfer of drug across the interfaces (capillary walls and cell membranes) can be by numerous transport mechanisms. The simplest is by Fickian (passive) diffusion, which is of the form:

$$FLUX_{a,b} = P.SA.(Cu_a - Cu_b)$$

where P is the permeability coefficient of the drug across the membrane, SA is the membrane surface area and Cu_a and Cu_b are the unbound concentration of drug on each side of the membrane.

When drugs can move freely from capillaries into interstitial fluid and tissue cells, i.e. capillary walls and cell membranes offer virtually no resistance to the movement of drug, the rate-limiting step controlling movement of drug into and out of an organ is the perfusion or blood flow to the organ.

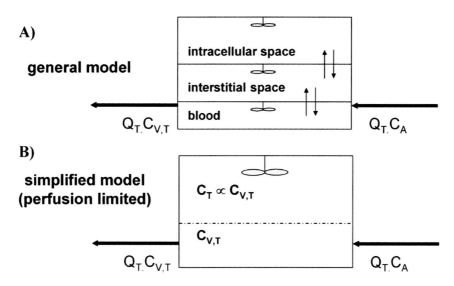

Figure 3. A general and simplified organ kinetic model.

Using this assumption the general model described above reduces to a single compartment. Known as the perfusion-limited or venous equilibration model (see Fig. 3B) the rate of change of amount of drug in the tissue is defined as:

$$\frac{dA_T}{dt} = Q_T.C_A - Q_T \cdot \frac{A_T}{V_T.K_p}$$

A_T is the mass of drug in the tissue, V_T is the anatomical volume of the tissue and K_p is the equilibrium tissue: plasma partition coefficient. As can be seen from the above equations the parameters can be separated into physiological/anatomical values and compound-dependent parameters. Values of tissue blood flows and anatomical volumes for many common laboratory animals and humans are available [15]. Thus, the key determinant, which requires estimation, is the K_p for each tissue in each species of interest. These values can be estimated using *in vivo*, *in vitro* and more recently *in silico* techniques [16, 17]. *In silico* methods are the most applicable to drug discovery and complement the high-throughput methodologies used by most pharmaceutical companies [17].

Metabolism

Drug metabolism is often the key route of elimination of the drug from the body. In general, the liver is the organ responsible for the metabolism of most drugs. Therefore, most pharmaceutical companies now routinely screen compounds for the metabolism properties using *in vitro* systems incorporating various animal and human liver preparations. To quantitatively estimate the rate of metabolism or clearance (CL) in humans *in vivo* the *in vitro* data must be extrapolated to the *in vivo* condition. A number of methodologies have been proposed in the literature and all require the use of mathematical models of the liver in order to make the extrapolation [18-22]. Such a methodology is shown in Fig. 4. A number of different liver models have been proposed including the venous equilibrium (well-stirred) model, the undistributed sinusoidal (parallel-tube) model and the dispersion model [23,24].

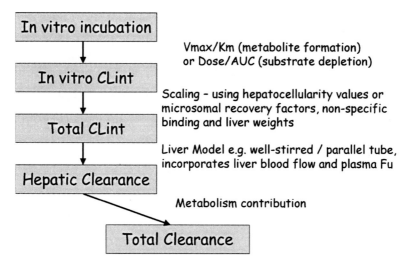

Figure 4. Liver metabolism *in vitro-in vivo* scale up methodology (adapted from Houston, 1994 [18]).

In theory, this approach can be applied to any tissue or organ which metabolizes the drug. However, limiting factors for expanding this approach to other tissues in a drug discovery setting are the availability of tissue or high-throughput assays and lack of quantitative information required to scale from the *in vitro* system.

Excretion

Other routes of elimination of a drug from the body include excretory processes in the kidney and into the bile from the liver. Filtration by the kidney has been well characterized and glomerular filtration rate can be used to estimate this part of renal clearance (C_{LR}) [25]. Although there has been considerable research in recent years on the active mechanisms in renal and biliary excretory processes, high-throughput assays and quantitative *in vitro-in vivo* correlations are currently lacking.

Integration - whole body models

Physiologically based pharmacokinetic models combine the models of absorption, distribution, metabolism and excretion in the attempt to predict the time-course of drug concentration in bodily fluids, normally blood or plasma, and the time-course of concentration in tissues [3-7]. An illustration of a PBPK model is shown in Fig. 5.

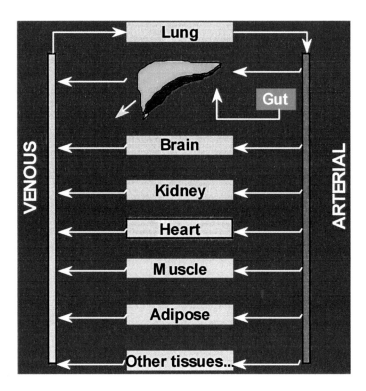

Figure 5. Schematic of a whole-body PBPK model.

Obviously every model is an abstraction or simplification and PBPK models are no exception. Normally, models of organs responsible for elimination of the drug are represented discretely in the whole body model. Large diffuse tissues such as fat, muscle and skin generally provide the largest storage capacity for drugs and so are also represented. Kinetic models of target organs are normally incorporated, while tissues that have negligible impact on the distribution of the drug are often ignored. All tissue models are connected together according to the design of the circulatory system.

Pharmacodynamics

Development of a model that will accurately simulate the pharmacokinetic system of a drug candidate is an achievement in its own right. With the addition of simple assumptions about therapeutic concentrations at the site of action such models can be used to give a first approximation of the dose-therapeutic response relationship. However, as the drug concentration (C)-effect (E) relationship is most frequently highly non-linear (see Fig. 6) [26], inclusion of pharmacodynamic models is often necessary to truly characterize the dose-response relationship. This is of particular importance when trying to make comparisons between drug-candidates in order to select the most suitable candidate for further development.

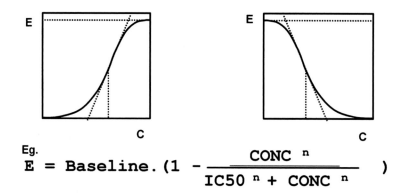

$$E = Baseline. (1 - \frac{CONC^n}{IC50^n + CONC^n})$$

Figure 6. Example of direct pharmacodynamic effects.

In the early stages of lead optimization when compounds are being tested for potency, ADME and physico-chemical properties and *in vitro* potency data (e.g. IC_{50} from a binding assay study) can be used as first approximation to define a pharmacodynamic relationship as per Fig. 6 [27, 28].

If such a model is combined with a PBPK model, quantified using the ADME and physico-chemical screening data, the likely dose-response, in terms of target (e.g. receptor or enzyme) inhibition or agonism at least, can be estimated. Thus, modelling and simulation can serve as an *in silico* screen utilizing the available data. Normally, as the number of suitable drug-candidates is reduced the remaining candidates are screened using *in vivo* animal pharmacology models. Often it is possible to establish pharmacodynamic relationships and models from the data obtained in these *in vivo* studies. Data can range from a simple *in vivo* confirmation of *in vitro* potency data, e.g. plasma enzyme inhibition, and more complex biomarkers of disease such as changes in hormone levels to complex behavioural changes in nervous system function. Models based on these data can be used in combination with PBPK models to first approximate the animal data (see Fig. 7) before attempting to extrapolate to humans.

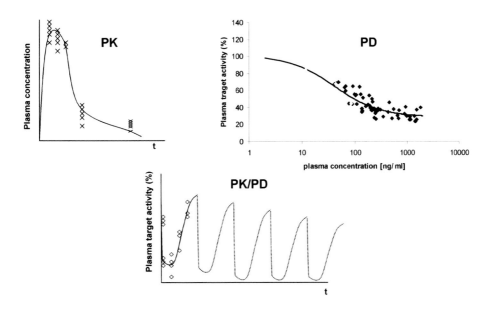

Figure 7. PK/PD Approach using *in vivo* animal data.

The advantages of using a modelling approach as opposed to using just the data, are that one can quantitatively incorporate independent data, any assumptions are explicated and models allow both interpolation and extrapolation beyond the conditions in which the experimental data were originally obtained. In particular, PBPK models can be used to estimate human kinetics from limited *in vitro* and *in silico* data.

The main uncertainty in such a combined PK/PD approach is scaling the pharmacodynamic model from the animal pharmacology model to the human situation. For drug targets that are situated in the blood this can be relatively straightforward. Although one cannot usually make predictions about final clinical endpoints, one can often extrapolate simple drug effects or biomarkers of clinical response. This becomes increasingly difficult when the target is not easily measurable even in animal models. It is also difficult when a link between responses in animal models and changes in disease state are not well understood. These uncertainties will decrease as more experience is obtained in the use of animal disease models, as our knowledge of human disease increases and as our current knowledge is applied quantitatively in an effort to predict human disease states and the therapeutic effects of drugs. A good example is the company Entelos Inc. [29, 30], which has developed sophisticated, mechanistic computer models of human physiology and disease. Models of this type can be used at many stages of the drug development process including drug discovery. When used in combination with predictive pharmacokinetic models, such as PBPK models, the dose-clinical therapeutic response of drug-candidates can be estimated long before the drug reaches the clinic.

CONCLUSION

Modelling and simulation based on the principles of pharmacokinetics and pharmacodynamics are useful tools in drug development. Biological system-based models allow one to separate the pharmacokinetic and pharmacodynamic characteristics of drug candidate into compound and biology-specific components. Advances in high-throughput experimental methods means that compound-specific data are generated for many drug-candidates at early stages of drug discovery. *In silico* tools have been developed to predict parameters that cannot be routinely measured. Biology-specific data are readily available in the scientific literature. With the advancement of sophisticated human physiology, disease and toxicity models the relationships between drug targets and clinical outcome are increasingly quantifiable. A combination of drug discovery data and computer methods allows scientists to explore the possibilities of novel drug targets or novel uses of existing targets at the beginning of a discovery project. In lead optimization one can discriminate between drug-candidates based on desirable pharmacokinetic properties and/or likely dose-effect characteristics in the target population.

Systematic use of these tools should lead to better clinical drug-candidates and a corresponding reduction in attrition during the far costlier clinical phases of drug development.

REFERENCES

[1]   DiMasi, J.A. (2001) Risks in new drug development: approval success rate for investigational drugs. *Clin. Pharmacol. Therapeut.* **69**:297-307.

[2]   Rowland, M., Thomas, T.N. (1988) *Clinical Pharmacokinetics. Concepts and Applications.* 2nd Edn, p.3. Lea and Febiger, Philadelphia.

[3]   Gerlowski, L.E., Jain, R.K. (1983) Physiologically based pharmacokinetic modeling: principles and applications. *J. Pharmaceut. Sci.* **72**:1103-1127.

[4]   Andersen, M.E. (1995) Development of physiologically based pharmacokinetic and physiologically-based pharmacodynamic models for applications in toxicology and risk assessment. *Toxicol. Letters* **79**:35-44.

[5]   Lutz, R.J., Dedrick, R.L., Zaharko, D.S. (1980) Physiological pharmacokinetics: an *in vivo* approach to membrane transport. *Pharmacol. Therapeut.* **11**:559-592.

[6]   MacDonald, A.J., Rostami-Hodjegan, A., Tucker, G.T., Linkens, D.A. (2002) Analysis of solvent central nervous system toxicity and ethanol interactions using a human population physiologically based kinetic and dynamic model. *Regulat. Toxicol. Pharmacol.* **35**:165-176.

[7]   Poulin, P., Theil, F.P. (2002) Prediction of pharmacokinetics prior to *In Vivo* studies. II. Generic physiologically based pharmacokinetic models of drug disposition. *J. Pharmaceut. Sci.* **91**:1358-1370.

[8]   Boxenbaum, B., D'Souza, R.W. (1990) Interspecies pharmacokinetic scaling, biological design and neoteny. In: *Advances in Drug Research*. (Testa, B., Ed.), pp. 139 -196. Academic Press, London.

[9]   Lin, J.H. (1995) Species similarities and differences in pharmacokinetics. *Drug Metab. Disposit.* **23**:1008-1021.

[10] Lavé, T., Coassolo, P., Reigner, B. (1999) Prediction of hepatic metabolic clearance based on interspecies allometric scaling techniques and *in vitro-in vivo* correlations. *Clin. Pharmacokinet.* **36**:211-231.

[11] Zuegge, J., Schneider, G., Coassolo, P., Lavé, T. (2001) Prediction of hepatic metabolic clearance in man - comparison and assessment of prediction models. *Clin. Pharmacokinet.* **40**:553-563.

[12] Agoram, B., Woltosz, W.S., Bolger, M.B. (2001) Predicting the impact of physiological and biochemical processes on oral drug bioavailability. *Adv. Drug Deliv. Rev.* **50**:S41-67.

[13] Grass, M., George, J., Sinko, P. (2002) Physiologically-based pharmacokinetic simulation modelling. *Adv. Drug Deliv. Rev.* **54**:433-451.

[14] Parrott, N., Lavé T. (2002) Prediction of intestinal absorption: comparative assessment of Gastroplus ™ and Idea ™. *Eur. J. Pharmaceut. Sci.* **17**:51-61.

[15] Brown, R.P., Delp, M.D., Lindstedt, S.L., Rhomberg, L.R., Beliles, R.P. (1997) Physiological parameter values for physiologically based pharmacokinetic models. *Toxicol. Indust. Hlth* **13**:407-484.

[16] Poulin, P., Theil, F.P. (2000) *A priori* prediction of tissue:plasma partition coefficients of drugs to facilitate the use of physiologically-based pharmacokinetic models in drug discovery. *J. Pharmaceut. Sci.* **89**:16-35.

[17] Poulin, P., Theil, F.P. (2002) Prediction of pharmacokinetics prior to *in vivo* studies. 1. Mechanism-based prediction of volume of distribution. *J. Pharmaceut. Sci.* **91**:129-156.

[18] Houston, J.B. (1994) Utility of in vitro drug metabolizing data in predicting *in vivo* metabolic clearance. *Biochem. Pharmacol.* **47**:1469-1479.

[19] Iwatsubo, T. *et al.* (1997) Prediction of *in vivo* drug metabolism in the human liver from *in vitro* metabolism data. *Pharmacol. Therapeut.* **73**:147-171.

[20] Lavé, T. *et al.* (1997) The use of human hepatocytes to select compounds based on their expected hepatic extraction ratios in humans. *Pharmaceut. Res.* **14**:152-155.

[21] Obach, R.S. *et al.* (1997) The prediction of human pharmacokinetic parameters from preclinical and *in vitro* metabolism data. *J. Pharmacol. Expl Therapeut.* **283**:46-58.

[22] Jones, H.M., Houston, J.B. (2004) Use of the substrate depletion approach for determining *in vitro* metabolic clearance: time dependencies in hepatocyte and microsomal incubations. *Drug Metab. Disposit.* **32**(9):973-982.

[23] Pang, K.S., Rowland, M. (1977) Hepatic clearance of drugs. I. Theoretical considerations of a "well-stirred" model and a "parallel tube" model. Influence of hepatic blood flow, plasma and blood cell binding, and the hepatocellular enzymatic activity on hepatic drug clearance. *J. Pharmacokinet. Biopharmaceut.* **5**:625-653 .

[24] Wilkinson, G.R. (1987) Clearance approaches in pharmacology. *Pharmacol. Rev.* **39**:1-47.

[25] Rowland, M., Thomas, T.N. (1988) *Clinical Pharmacokinetics. Concepts and Applications.* 2nd Edn, pp. 160-163. Lea and Febiger, Philadelphia.

[26] Gabrielsson, J., Weiner, D. (2000) *Pharmacokinetic and Pharmacodynamic Data Analysis: Concepts and Applications.* 3rd Edn, pp.175-257. Swedish Pharmaceutical Press, Stockholm.

[27] Visser, S.A.G. *et al.* (2003) Mechanism-based pharmacokinetic/pharmacodynamic modeling of the electroencephalogram effects of GABAA receptor modulators: *in vitro-in vivo* correlations. *J. Pharmacol. Expl Therapeut.* **304**:88-101.

[28] Gomeni, R., Bani, M., D'Angeli, C., Corsi, M., Bye, A. (2001) Computer-assisted drug development (CADD): an emerging technology for designing first-time-in-man and proof-of-concept studies from preclinical experiments. *Eur. J. Pharmaceut. Sci.* **13**:261-270.

[29] Musante, C.J., Lewis, A.K., Hall, K. (2002) Small- and large-scale biosimulation applied to drug discovery and development. *Drug Discov. Today* **7**:s192-s196.

[30] Kansal, A.R. (2004) Modeling approaches to type 2 diabetes. *Diabetes Technol. Therapeut.* **6**:39-47.

THE FINAL CURTAIN

KEVIN DAVIES

Editor-in-Chief, Bio-IT World, 1 Speen Street,
Framingham MA 01701, U.S.A.
E-Mail: Kevin_davies@bio-itworld.com

Received: 18th August 2004 / Published: 22nd July 2005

The 2004 Beilstein Bozen Workshop succeeded admirably in its stated goal of bringing together a diverse collection of interdisciplinary researchers, resulting in a stimulating and provocative conference. Despite their broad range of backgrounds - chemistry and biology, mathematics and biophysics, academia and big pharma - the speakers and attendees were united in their desire to understand more fully the dynamic interaction of genes, proteins and other cellular constituents, whether it be to model cellular pathways, or design and synthesize improved libraries of small-molecule drugs.

The preceding chapters in this monograph illustrate the remarkable quality of the presentations at the 2004 meeting. In this concluding chapter, I will simply aim to put some of these advances into a broader context.

THE PROMISE OF CHEMICAL GENOMCIS

In April 2003, the NIH marked the successful conclusion of the Human Genome Project - a massive international endeavour that was completed ahead of schedule, and under budget [1]. The full sequence is publicly available via several web sites, such as the Golden Path at UC Santa Cruz (genome.ucsc.edu). By May 2004, the month of the symposium, the sequence of approximately half of the human chromosomes had been published, along with the complete genomes of some 200 species, including mouse, rat, as well as preliminary drafts of the dog and chimpanzee genomes. Shortly after the completion of the human genome sequence, Francis Collins and his colleagues published a manifesto for the National Human Genome Research Institute, laying out a dozen or more key strategic goals over the next several years [2].

Interestingly, one of those goals encouraged academic groups to play a bigger role in drug discovery, using the tools of chemical genomics. Collins *et al.* wrote:

> "[Gleevec] offers promise that therapies based on genomic information will be particularly effective... A promising example of the gene-based approach to therapeutics is 'chemical genomics.' Providing such access more broadly... could lead to the discovery of a host of probes for biological pathways... Also needed are more powerful technologies for generating deep molecular libraries... A centralized database of screening results should lead to further important biological insights... Generating molecular probes for exploring the basic biology of health and disease in academic laboratories would not supplant the major role of biopharmaceutical companies in drug development, but could contribute to the start of the pipeline." [2].

Of course, academia has a rich tradition in drug development. Erythropoietin, tamoxifen, Taxol, AZT and many other important drugs were either purified or extracted by academic researchers, or required the tireless lobbying of physicians in order to gain the attention of drug companies [3]. The NIH has launched a Chemical Genomics Center, the flagstone of a ten-center nationwide Molecular Libraries Screening Centers Network.

The importance of chemical biology and improved synthetic techniques to revitalize the drug discovery process (and basic research) was one of the central themes of the symposium. Steve Ley reminded the audience that there is, by some estimates, a universe of 1060 small molecules. The Chemical Abstracts Service registry lists more than 23 million organic and inorganic compounds. And the famous Beilstein database, a compendium of organic chemicals going back to 1771, includes records of some 9.3 million compounds.

By contrast, the number of therapeutic targets is remarkably small. According to several investigators, the total number of targets in the 'druggable genome' amounts only to about 3,000 - that is, the number of genes in the human genome that encode biochemically tractable targets of medical/disease relevance [4](Fig. 1).

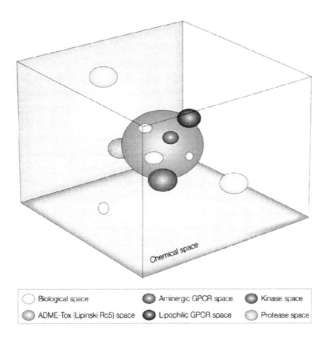

Figure 1. Chemical space and the druggable genome. The box depicts the complete set of all possible molecular structures, or chemical space - 1060 or more. The coloured spheres represent subsets of potentially therapeutic molecules. (Published with permission by Nature Publishing Group. Peti-Zeman, S. (2004) Exploring biological space. *Nature Reviews Drug Discovery*, Horizon Symposia; May).

Given the vast excess of small molecules compared to therapeutic targets, why has the production of new medical entities declined so dramatically in recent years? There are many factors (see below), but part of the problem is at the lead identification stage. Many mediocre small-molecule libraries have been created during the past 10-20 years by decorating basic core structures, rather than building diversity into the basic core structure. (Steve Ley's description of this depressing trend was "absolutely pathetic.") Ley's group takes a target-oriented synthesis approach to facilitate the synthesis of important compounds (Fig. 2).

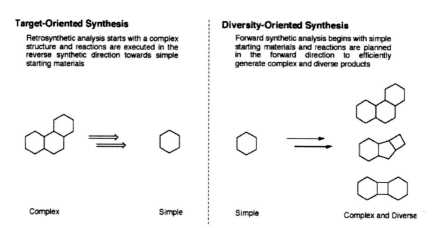

Target-Oriented Synthesis

Retrosynthetic analysis starts with a complex structure and reactions are executed in the reverse synthetic direction towards simple starting materials

Diversity-Oriented Synthesis

Forward synthetic analysis begins with simple starting materials and reactions are planned in the forward direction to efficiently generate complex and diverse products

Complex   Simple   Simple   Complex and Diverse

Figure 2. Target-oriented synthesis and diversity-oriented synthesis. (Published with permission from Cell Press. Burk, M.D., Lalic, G. (2002) *Chemistry & Biology* 9:535-541).

A similar theme was sounded by Keith Russell (AstraZeneca). It is not uncommon for drug screens employing in-house or commercial libraries of a million compounds or more to fail, at least initially, because of lack of diversity, quality and stability of the molecules. Both presenters underscored the importance of new software and robotic devices to improve the synthesis of new compound libraries.

The process of diversity-oriented synthesis (Fig. 2), championed by chemists such as Harvard's Stuart Schreiber, shows how chemical diversity can be injected into libraries using just a handful of combinatorial steps [5]. Such methods not only bode well for lead identification in the future, but also provide an arsenal of promising new tools for studies of chemical genetics, whether it be the incorporation of novel amino acids into proteins, the combinatorial synthesis of novel antibiotics, or the development of *in vivo* biological assays (Fig. 3) [6]. The explosion of public funding in this area, and new database resources such as PubChem and Chembank, auger well for the future development of this field.

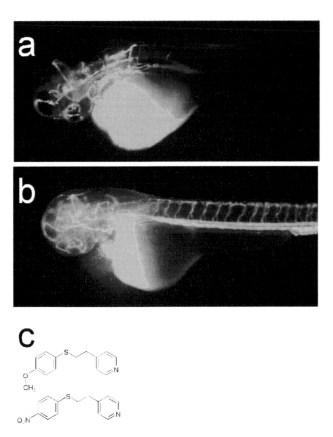

Figure 3. Chemical rescue of zebrafish genetic cardiovascular defect. Microangiograms of a gridlock zebrafish embryo before **(a)** and after **(b)** treatment with the small molecule GS4021. **(c)** Shows the chemical structures of two small molecules gridlock suppressors identified by *in vivo* screening, with GS4021 on the top. (Published with permission from Nature Publishing Group. Peterson, R.T. *et al.* (2004) *Nature Biotechnology* 22:595-599).

Several speakers addressed the importance of virtual screening and biosimulation for modelling drug-protein interactions and 'reducing the haystack' of potential drug candidates, and the development of new software algorithms and databases to facilitate research advancement. Sometimes, however, the answer is to make the haystack bigger. As described in the 2002 Beilstein workshop report, Graham Richards and his colleagues' distributed computing project produced a virtual library of 3.5 billion molecules, by concocting 100 modifications for each of 35 million compounds, with considerable success [7].

Taken together, these initiatives will help investigators tackle one of the most challenging problems faced in medicine today - the crippling cost and time required for drug discovery.

THE COST OF DRUG DISCOVERY

The importance of boosting the quality and speed of discovery of new drugs cannot be overestimated. The net cost of drug development is put at $800 million, according to an oft-quoted 2001 study from the Tufts Center for Drug Development. (This includes the 'cost of capital,' the result of treating R&D costs as an investment rather than an expense.)

In his book *The $800 Million Pill*, author Merrill Goozner points out that this $800 million estimate also includes the cost of developing me-too drugs; the cost of developing stereoisomers of approved drugs to extend a drug patent; and the practice of conducting clinical trials not for the purpose of seeking FDA approval, but simply to gather data to persuade doctors to prescribe a particular brand of medication. "In short," writes Goozner, "if the industry funded academic economists at Tufts had factored out the half of industry research that is more properly categorized as corporate waste, their number would have been similar to that of the Global Alliance." [3].

A few years ago, however, the Global Alliance for Tuberculosis Drug Development convened a panel of industry experts to calculate the cost of developing a new drug against tuberculosis. The report, published in October 2001 [8], concluded that the total costs to discover and develop a new anti-TB drug - including the costs of failures -ranges from $115 million to $240 million. These costs could be divided into $40-125 million for discovery efforts, and $76-115 for preclinical development through phase III trials.

Even drugs that are approved in record time, such as Novartis' Gleevec, which was approved by the FDA in 2001 in a mere 10 weeks, took a decade to reach the market. The molecule formerly known as STI-571 was originally synthesized by Jurg Zimmerman in 1992. Much of the increase in drug development time and cost during the past two decades is in the area of clinical trials - difficulty recruiting patients, meeting FDA guidelines, and the more complex nature of the diseases being treated all conspire to lengthen the process and raise the cost. Clearly, greater insight earlier in the process as to the likely efficacy and toxicity of potential new drugs could reap profound dividends later on.

The recent saga of Iressa is instructive. The lung cancer drug was approved after a costly trial involving some 12,000 patients, which caused the company something of a dilemma, as indicated by Sir Tom McKillop, CEO of AstraZeneca: "There are 50% of patients with lung cancer who do not get any benefit from Iressa ... but then there is a small group, maybe 10-20% of patients [who] get an almost a miraculous response, and their whole life can be transformed and extended for years." [9].

The reason for these contrasting results has now been clarified. In a pair of papers published in 2004, researchers in Boston found that patients who respond to Iressa have acquired mutations in the target protein that increase the binding of the drug. A new diagnostic test is currently in preparation [10,11].

The examples of Gleevec and Iressa, two of the most high-profile cancer drugs approved in recent years, demonstrate how critical it is to marry rational drug design with improved understanding of the target patient population. Most, perhaps all, drugs do not produce the same results in all patients. Using pharmacogenomic and other tools to stratify the patient population so that the positive effects of a drug are not shrouded by non-responders or worse, adverse events, will pay dividends in the long run.

IN CONCLUSION

The convergence of technologies and disciplines showcased at the 2004 Beilstein symposium invited several speakers to talk about the new holy grail of modern biology - systems biology. In 2003, researchers at Curagen published a proteome map of some 7,000 proteins in *Drosophila* that they dubbed "The dawn of systems biology" [12]. Be that as it may, the explosion of interest in integrative biology - the interplay of genome-wide expression, proteomic, metabolomic and *in silico* studies - lends credence to the view that we are on the verge of describing in more holistic terms the functioning of cellular systems.

Several speakers at the symposium discussed areas in which this could be achieved, including work at Hoffman-LaRoche, in collaboration with Entelos Inc., to produce 'virtual patients' for diseases such as asthma, obesity and arthritis [13]. In a related vein, researchers at Gene Network Sciences, in Ithaca New York, are building a computational model of a colon cancer cell.

258

It is to be hoped that the rich potential inherent in these models acquires sufficient validation from the academic community that they might form an integral part of the discussion at the 2006 Beilstein Bozen Workshop.

REFERENCES

[1]   Davies, K. (2001) *Cracking the Genome*. The Free Press, New York.

[2]   Collins, F.S., Green, E.D., Guttmacher, A.E., Guyer, M.S. (2003) A vision for the future of genomics research. *Nature* **422**:835-847.

[3]   Goozner, M. (2004) *The $800 Million Pill*. UC Berkeley Press, Berkeley.

[4]   Hopkins, A.L., Groom, C.R. (2002) The druggable genome. *Nature Rev. Drug Discov.* **1**:727-730.

[5]   Burke, M.D., Berger, E.M., Schreiber, S.L. (2003) Generating diverse skeletons of small molecules combinatorially. *Science* **302**:613-618.

[6]   Peterson, R.T., Shaw, S.Y., Peterson, T.A. *et al.* (2004) Chemical suppression of a genetic mutation in a zebrafish model of aortic coarctation. *Nature Biotech.* **22**:595-599.

[7]   Richards, W.G. (2003) Pattern recognition and distributed computing in drug design. In: *Molecular Informatics: Confronting Complexity*. (Hicks, M. G., Kettner, C. Eds), Proceedings of the Beilstein-Institut Workshop, May 13-16, 2002, Bozen, Italy.

[8]   Global Alliance for Tuberculosis. The economics of TB drug development. (2001). http://www.tballiance.org/

[9]   Humphreys, A. (2003) Ability to manage. *Med. Ad. News* **22**:36-49.

[10]   Lynch, T.J., Bell, D.W., Sordella, R. *et al.* (2004) Activating mutations in the epidermal growth factor receptor underlying responsiveness of non-small-cell lung cancer to gefitinib. *New Engl. J. Med.* **350**:2129-2139.

[11]   Paez, J.G., Janne, P.A., Lee, J.C. *et al.* (2004) EGFR mutations in lung cancer: Correlation with clinical response to gefitinib therapy. *Science* **304**:1497-1500.

[13]   Giot, L., Bader, J.S., Brouwer, C. *et al.* (2003) A protein interactive map of *Drosophila melanogaster*. *Science* 302:1727-1736.

[14]   Uehling, M.D. (2003) Model patient. *Bio-IT World* **2**:30-36.

BIOGRAPHIES

Konrad Bleicher

09/85-02/95	PhD Organic Chemistry, Tübingen University; Prof. E. Bayer
10/00-today	Hoffmann-La Roche, Basel, Switzerland; Scientific Specialist, Head of Combinatorial Chemistry
04/99-09/00	Hoffmann-La Roche, Basel, Switzerland; Senior Scientist, Member of New Lead Chemistry Initiative Responsibility to build up a high throughput synthesis and purification platform for the generation of large compound libraries
10/97-04/99	Hoffmann-La Roche, Basel, Switzerland; Research Scientist Parallel Synthesis Group CNS Support of various CNS projects using parallel synthesis techniques
09/96-11/97	Novartis, East Hanover, USA; Dr. J. Wareing Postdoctoral Fellow Combinatorial Chemistry Resin and Linker synthesis for solid supported combinatorial chemistry
04/95-09/96	Sandoz, Basel, Switzerland; Dr. R.Giger Postdoctoral Fellow Combinatorial Chemistry Solid supported synthesis of compound collections using Chiron's multi-tipin technology

Konrad Bleicher is author of ca. 25 patents and scientific publications.

Virginia Cornish

graduated summa cum laude from Columbia University with a B.A. in Biochemistry in 1991, where she did undergraduate research with Professor Ronald Breslow in the Chemistry Department. She then moved west to do research with Professor Peter Schultz in the Chemistry Department at the University of California at Berkeley as an NSF Predoctoral Fellow. In Professor Schultz's laboratory she helped develop a new methodology for incorporating synthetic amino acids into proteins using the protein biosynthetic machinery. In 1996, she became an NSF Postdoctoral Fellow in the Biology Department at M.I.T. under the guidance of Professor Robert Sauer. At M.I.T. she initiated an independent project that is the basis for the directed evolution program in her laboratory at Columbia. Virginia joined the Chemistry Department at Columbia in 1999, working at the interface of chemistry and biology.

Her laboratory brings together modern methods in synthetic chemistry and DNA technology to co-opt biological systems for the synthesis of new materials, understanding the function of these systems by challenging their specificity at the molecular level. Her research has been recognized by numerous awards including a Sloan Foundation Fellowship, a Beckman Young Investigator Award, and a NSF Career Award.

Biographies

Athel Cornish-Bowden

carried out his undergraduate studies at Oxford, obtaining his doctorate with Jeremy R. Knowles in 1967. After three post-doctoral years in the laboratory of Daniel E. Koshland, Jr., at the University of California, Berkeley, he spent 16 years as Lecturer, and later Senior Lecturer, in the Department of Biochemistry at the University of Birmingham. Since 1987 he has been Directeur de Recherche in three different laboratories of the CNRS at Marseilles. Although he started his career in a department of organic chemistry virtually all of his research has been in biochemistry, with particular reference to enzymes, including pepsin, mammalian hexokinases and enzymes involved in electron transfer in bacteria. He has written several books relating to enzyme kinetics, including *Analysis of Enzyme Kinetic Data* (Oxford University Press, 1995) and *Fundamentals of Enzyme Kinetics* (3rd edition, Portland Press, 2004). Since moving to Marseilles he has been particularly interested in multi-enzyme systems, including the regulation of metabolic pathways. More generally, he has long had an interest in biochemical aspects of evolution, and his semi-popular book in this field, *The Pursuit of Perfection*, will be published by Oxford University Press in 2004.

Kevin Davies

is the Editor-in-Chief of *BioIT World*, the monthly magazine covering information technology and life sciences published by IDG. He is the author of Cracking the Genome, an accessible and comprehensive account of the conclusion of the Human Genome Project.

Davies graduated from Oxford University and obtained in Ph.D in genetics from the University of London. After conducting postdoctoral research at MIT and Harvard Medical School, he joined the editorial staff of the prestigious British science journal *Nature* in 1990. In 1992, Kevin founded *Nature Genetics*, the world's leading genetics and genomics journal. He later served as the science editor at the Howard Hughes Medical Institute in Chevy Chase, Maryland, the largest medical philanthropy in the United States. Prior to launching *BioIT World*, Davies served as editor-in-chief of Cell Press.

Davies' latest book, Cracking the Genome, has been translated into 15 languages. He is also the author (with Michael White) of *Breakthrough: The Race to Find the Breast Cancer Gene*.

Ernesto Freire

is the Henry Walters Professor in the Department of Biology at The Johns Hopkins University. He is also Professor in the Department of Biophysics. Dr. Freire has been at Johns Hopkins since 1986. He obtained his Ph.D. in Biophysics at the University of Virginia. Dr. Freire has served in numerous scientific boards, scientific advisory committees, editorial boards, and is a member of the Biophysical Society, American Chemical Society, American Association for the Advancement of Science and Protein Society.

Dr. Freire has pioneered the application of thermodynamic methods to the development of new molecular design algorithms aimed at engineering drugs that exhibit extremely high affinity, selectivity and maintain their effectiveness in the face of mutations causing drug resistance and genomic diversity. Dr. Freire has authored over 170 scientific publications and holds several patents.

Johann Gasteiger

(Johnny), studied chemistry at the University of Munich, ETH and the University of Zürich and received his Doctorate in Organic Chemistry from the University of Munich in 1971. Following a postdoctoral fellowship at the University of California in Berkeley in 1971-1972 Johnny taught at the Technical University of Munich. In 1994 he moved to the University of Erlangen-Nuremberg where he co-founded the "Computer-Chemie-Centrum". He is one of the founders of Chemoinformatics in Germany and has produced more than 250 scientific publications in this field Johnny was consultant to the Beilstein Institute and to FIZ CHEMIE BERLIN, where he was the project manager for the development of the ChemInformRX reaction database. He is a past chairman (1994-1996) of the Division "Chemistry-Information-Computer" of the German Chemical Society, and has served as Vice-Chairman of the Working Party "Computational Chemistry" of the Federation of European Chemical Societies (1986-2002). In 1991 Johnny was awarded with the Gmelin-Beilstein Medal of the German Chemical Society for Achievements in Computer Chemistry and in 1997 he received the Herman-Skolnik-Award of the Division of Chemical Information of the American Chemical Society.

Richard Goldstein

obtained his Ph.D. using experimental and computational methods to study electron transfer in bacterial photosythesis. After a brief stay teaching Physics in China, he worked with Peter Wolynes developing methods to predict protein tertiary structures. He then spent eight years on the faculty at the University of Michigan. He left academia briefly to become Head of Bioinformatics at Siena Biotech, and then moved to the Mathematical Biology Division at the National Institute for Medical Research in Mill Hill, London. His research focuses on the relationship between a protein's structure, function, and other properties and the evolutionary processes through which these properties emerged. These efforts have included methods of identifying and aligning distant protein homologs, examining the evolutionary record of related sets of proteins in order to determine characteristics of specific proteins, developing better models for phylogenetic reconstruction, and using simplified theoretical and computational models to develop deeper insights into how proteins can be understood in their evolutionary context.

Martin G. Hicks

is a member of the board of management of the Beilstein-Institut. He received an honours degree in chemistry from Keele University in 1979. There, he also obtained his PhD in 1983 studying synthetic approaches to pyridotropones under the supervision of Gurnos Jones. He then went to the University of Wuppertal as a postdoctoral fellow, where he carried out research with Walter Thiel on semi-empirical quantum chemical methods. In 1985, he joined the computer department of the Beilstein-Institut where he worked on the Beilstein Database project. His subsequent activities involved the development of cheminformatics tools in the areas of substructure searching and reaction databases, and products such as Current Facts and CrossFire. After brief sojourns as the managing director of the Beilstein Verlagsgesellschaft in 1997 and subsequently the Beilstein GmbH from 1998 - 2000, he returned home to the Beilstein-Institut as head of the funding department in 2000.

He is particularly interested in furthering interdisciplinary communication between chemistry and neighbouring scientific areas and has been organizing the Beilstein Bozen Workshops since 1988.

Carsten Kettner

studied biology at the University of Bonn and obtained his diploma at the University of Göttingen in the group of Prof. Gradmann which had the pioneering and futuristic name - "Molecular Electrobiology". This group consisted of people carrying out research in electrophysiology and molecular biology in fruitful cooperation. In this mixed environment, he studied transport characteristics of the yeast plasma membrane using patch clamp techniques. In 1996 he joined the group of Dr. Adam Bertl at the University of Karlsruhe and undertook research on another yeast membrane type. During this period, he successfully narrowed the gap between the biochemical and genetic properties, and the biophysical comprehension of the vacuolar proton-translocating ATP-hydrolase. He was awarded his Ph.D for this work in 1999. As a post-doctoral student he continued both the studies on the biophysical properties of the pump and investigated the kinetics and regulation of the dominant plasma membrane potassium channel (TOK1). In 2000 he moved to the Beilstein-Institut to represent the biological section of the funding department. Here, he is responsible for the organization of symposia (sic!), administering research (proposals) and funding, as well as, the development of new projects and products for the Beilstein-Institut.

Hugo Kubinyi

studied chemistry in Vienna, Austria. After his Ph.D. thesis at the Max Planck Institute of Biochemistry in Munich he continued as a PostDoc at the German Cancer Research Centre in Heidelberg. In 1966 he joined Knoll AG, later a subsidiary of BASF AG, and in 1985 he moved to BASF AG. Since 1987, until his retirement in summer 2001, he was responsible for the Molecular Modelling, X-ray Crystallography and Drug Design group of BASF, since early 1998 also for Combinatorial Chemistry in the Life Sciences.

He is Professor of Pharmaceutical Chemistry at the University of Heidelberg, former Chair of The QSAR and Modelling Society, and IUPAC Fellow. From his scientific work resulted five books on QSAR, 3D QSAR, and Drug Design (the German book "Wirkstoffdesign" received the 1999 Book Award of the FCI, Association of Chemical Industry) and about 90 publications. He is a member of several Scientific Advisory Boards, coeditor of the Wiley-VCH book series "Methods and Principles in Medicinal Chemistry", and member of the Editorial Boards of several scientific journals.

Steve Ley

is BP (1702) Professor of Organic Chemistry at the University of Cambridge, and Fellow of Trinity College. He is Immediate Past President of the Royal Society of Chemistry and was made a CBE in January 2002.

He was appointed to the staff at Imperial College in 1975 and became head of department in 1989. He was elected to the Royal Society (London) in 1990, and came to Cambridge in 1992.

Steve's research involves the discovery and development of new synthetic methods and their application to biologically active systems. His group has published extensively on the synthesis of natural products and to date more than 100 target compounds have been synthesised. The group is also developing new ways of making complex carbohydrates and developing new strategies for combinatorial chemistry. Steve Ley's work of 530 papers has been recognised by 14 major prizes and awards, the most recent of which was the Ernest Guenther Award from the American Chemical Society.

Thomas S. Leyh

received a Ph. D. in biophysics from the University of Pennsylvania in 1983. He joined the faculty at the Albert Einstein College of Medicine in New York in 1989, where he is currently a Professor of Biochemistry. Prof. Leyh is a mechanistic enzymologist with a long-standing interest in sulfur biochemistry, GTPase function, and the conformational coupling of energetics. His group has recently demonstrated that enzymes in the cysteine biosynthetic pathway self-organize into a multifunctional protein complex out of which emerges new catalytic function that orchestrates the activities of the complex. John Andreassi, Ph. D., is a postdoctoral fellow working with Dr. Leyh to initiate the genomic enzymology program.

Alex MacDonald

is a Ph.D pharmacokineticist and bioengineer. He is currently working as a senior scientist at Roche Pharma Research in Basel, Switzerland, in the non-clinical modelling and simulation group. His expertise in physiologically-based pharmacokinetic and pharmacodynamic modelling and pre-clinical pharmacokinetics. His previous experience includes modelling and simulation scientist positions at Novartis Pharma AG and with the UK government. Prior to gaining his Ph.D Alex worked for a number of years as a control systems design engineer in the automotive industry.

Biographies

Jason Micklefield

received a PhD in Organic Chemistry from the University of Cambridge in 1993 working with Prof Sir Alan R. Battersby to complete the first total synthesis of Haem d1. This was followed by a NATO postdoctoral fellowship at the University of Washington in Seattle, USA, with Prof Heinz G. Floss, investigating enzyme mechanisms. In 1995 he became a Lecturer in Organic Chemistry at Birkbeck College, University of London, before moving to the University of Manchester Institute of Science and Technology in 1998 where he is now a Reader in Organic and Biological Chemistry. His current research interests are at the chemistry-biology interface and include the synthesis, conformational analysis and biophysical evaluation of modified nucleic acids and oligonucleotide mimics. His other research programmes are concerned with engineering the biosynthesis of nonribosomal peptide secondary metabolites and the use of biotransformations in synthesis.

Keith Russell

was born in the UK and obtained his PhD from the University of Cambridge, UK with Dr. A. B. Holmes in 1984 in the area of synthetic organic chemistry. He was then awarded a NATO fellowship to do post doctoral studies with Professor L. A. Paquette at Ohio State University (1984-1986). In 1986, he took up a position as a Senior Research Chemist in the Medicinal Chemistry Department of AstraZeneca Pharmaceuticals in Alderley Park. In 1989 he transferred to the Medicinal Chemistry Department in the Wilmington, DE research site of AstraZeneca Pharmaceuticals in the USA, where he is now Director of Chemistry in the CNS Discovery Area. Keith has worked in a number of areas of medicinal chemistry. He was a key player in the team that discovered ZD6169 and ZD0947, the first bladder selective potassium channel openers aimed at urge urinary incontinence. He led the Tachykinin team to deliver a selective N_{K1} antagonists for depression, a dual $N_{K1}N_{K2}$ antagonist for asthma, as well as the early stages of the NK2 antagonist project aimed at urge incontinence. He has contributed to several development teams and co-led a Global Tackykinin R&D Project aimed at increasing the efficiency of the R to D transition process. Keith has been a major force behind the introduction and application of automation, intranet, combinatorial chemistry and other new chemical technologies into AstraZeneca drug discovery. More recently, Keith has spearheaded an initiative in "Chemical Genetics" aimed at more fully integrating chemistry activities throughout the entire discovery value chain. An ACS and AAAS member, he is the author of over 30 peer-reviewed publications and is an inventor on over 25 patents emanating from his work at AstraZeneca. Keith is a reviewer for Journal of the American Chemical Society and Bioorganic and Medicinal Chemistry Letters.

Gisbert Schneider

born 1965 in Fulda, Germany; studied biochemistry and computer science at the Free University (FU) in Berlin; 1994, PhD in bioinformatics on neural networks and evolutionary algorithms; post-doctoral work on peptide design (with Prof. Wrede, FU Berlin), protein folding simulation (with Prof. Schimmel, M.I.T., Cambridge, USA), analysis of protein targeting signals (with Prof. von Heijne, University of Stockholm, Sweden) and prediction of membrane protein topology (with Prof. Schulten, Max-Planck-Institut Frankfurt, Germany); 1997-2002

F.Hoffmann-La Roche AG, Basel, Switzerland, head of cheminformatics; scientific research on combinatorial drug design, virtual screening, and genome analysis. Current position: Beilstein Professor of Cheminformatics at Johann Wolfgang Goethe-Universität, Frankfurt; research focus on adaptive systems in molecular design.

Nicholas J. Westwood

carried out his doctoral studies in chemical biology at Oxford University under the supervision of Professor Christopher Schofield (1992-1995). He then worked with Professor Philip Magnus FRS as a NATO post-doctoral fellow on synthetic approaches to Taxol (1995-1998), and spent a further 3 years at the Institute of Chemistry and Cell Biology (ICCB) at Harvard Medical School (1998-2001). In Professor Matthew Shair's laboratory, he prepared a collection of 2000+ compounds related in structure to the natural product, galanthamine using solid phase synthesis technology. One member of this library, named secramine, is currently being used to study protein trafficking. He also became interested in high throughput screening technologies during his time in Professor Timothy Mitchison's group at ICCB. A particular focus has been on chemical genetic approaches in cellular microbiology (in a continuing collaboration with Professor Gary Ward at the University of Vermont).

Nick was appointed to a Royal Society University Research Fellowship and lecturer position in Chemical Genetics at the University of St Andrews in 2001. The main research interests of his laboratory remain at the interface of chemistry and biology. His group uses a range of methods in synthetic organic chemistry to address fundamental biological questions. Projects include the chemical characterisation and optimisation of the small molecule tool, blebbistatin, synthetic programmes in diversity oriented synthesis and studies aimed at protein target identification. He has also recently established a HTS facility at St Andrews. He retains an interest in all high throughput synthesis technology and will be on secondment to Pfizer later this year in an EPSRC-funded scheme to improve industrial-academic communication in this research area.

Peter Willett

obtained an Honours degree in Chemistry from Exeter College, Oxford in 1975 and then went to the Department of Information Studies, University of Sheffield where he obtained an MSc in Information Studies. Following doctoral and post-doctoral research on computer techniques for the processing of databases of chemical reactions, he joined the staff of the University of Sheffield as a Lecturer in Information Science in 1979. He was awarded a Personal Chair in 1991 and a DSc in 1997, and is now the Head of the Department. He is a Fellow of the Chartered Institute of Library and Information Professionals, and was the recipient of the 1993 Skolnik Award of the American Chemical Society, of the 1997 Distinguished Lecturer Award of the New Jersey Chapter of the American Society for Information Science, of the 2001 Kent Award of the Institute of Information Scientists, and of the 2002 Lynch Award of the Chemical Structure Association Trust. He is included in Whos Who, is a member of the editorial boards of four international journals, and has been involved in the organisation of many national and international conferences in various aspects of information retrieval. Professor Willett heads a large research group studying novel computational techniques for the processing of chemical and biological information, and has over 400 publications describing this work.

His current interests include: database applications of cluster analysis, evolutionary computing and graph theory; molecular similarity and molecular diversity analysis; the comparison of chemical and biological 3D structures; and the use of citation data for the evaluation of academic research performance.

INDEX

tubulin depolymerizer 151
tumour suppressor 14

V

vancomycin 88
virtual
       library 140
       ligand 208
       screening 53, 63, 119, 140, 198
       screening method 120

Y

yeast three-hybrid assay 104

Z

zanamivir 60